Problem Books in Mathematics

Edited by P. Winkler

Problem Books in Mathematics

Series Editors: Peter Winkler

(continued after subject index)

Wolfgang Schwarz

40 Puzzles and Problems in Probability and Mathematical Statistics

 Springer

Wolfgang Schwarz
Universität Potsdam
Humanwissenschaftliche Fakultät
Karl-Liebknecht Strasse 24/25
D-14476 Potsdam-Golm
Germany
wschwarz@uni-potsdam.de

Series Editor:
Peter Winkler
Department of Mathematics
Dartmouth College
Hanover, NH 03755
USA
Peter.winkler@dartmouth.edu

ISBN-13: 978-1-4419-2522-0   e-ISBN-13: 978-0-387-73512-2

Mathematics Subject Classification (2000): 60-xx

We often felt that there is not less, and perhaps even more, beauty in the result of analysis than there is to be found in mere contemplation.

Niko Tinbergen, *Curious Naturalists* (1958).

Preface

As a student I discovered in our library a thin booklet by Frederick Mosteller entitled *50 Challenging Problems in Probability*. It referred to a supplementary "regular textbook" by William Feller, *An Introduction to Probability Theory and its Applications*. So I took this one along, too, and started on the first of Mosteller's problems on the train riding home. From that evening, I caught on to probability. These two books were not primarily about abstract formalisms but rather about basic modeling ideas and about ways — often extremely elegant ones — to apply those notions to a surprising variety of empirical phenomena. Essentially, these books taught the reader the skill to "think probabilistically" and to apply simple probability models to real-world problems.

The present book is in this tradition; it is based on the view that those cognitive skills are best acquired by solving challenging, nonstandard probability problems. My own experience, both in learning and in teaching, is that challenging problems often help to develop, and to sharpen, our probabilistic intuition much better than plain-style deductions from abstract concepts.

The problems I selected fall into two broad categories, even though it is in the spirit of this collection that there actually is no sharp dividing line between them. Problems related to probability theory come first, followed by problems related to the application of probability in the field of mathematical statistics. Statistical applications are by now probably the most important reason to study probability. Thus, it seemed important to me to select problems illustrating in an elementary but nontrivial way that probabilistic techniques form the basis of many statistical applications. All problems seek to convey a nonstandard aspect or an approach that is not immediately obvious.

The word *puzzles* in the title refers to questions in which some qualitative nontechnical insight is most important. Ideally, puzzles can teach a productive new way of framing or representing a given situation. An obvious example of this is the introductory "To Begin Or Not To Begin." Although the border between the two is not perfectly sharp (and crosses occur), *problems* tend to require a more systematic application of formal tools and to stress more technical aspects. Examples of this deal with the distribution of various statistics such as the range or the maximum, the linearization of problems with the powerful so-called Delta technique, or the precision (standard error) of certain estimates. These examples have in common that they all seek to apply basic probability notions to more or less applied situations. Such approaches — the Delta technique mentioned earlier is a typical example — are not usually covered in introductory books. Rather, they often appear hidden behind thick technical jargon in more advanced treatments, even though their probabilistic background is often essentially elementary. Thus, a major aim of the present collection is to help bridge the wide gap between introductory texts and rigorous state-of-the-art books.

Many puzzles and problems presented here are either new within a problem-solving context (although as topics in fundamental research they are of course long known) or variations of classical problems. A small number of particularly instructive problems have been taken from previous sources, which in this case are generally given (the titles referenced at the end of the book contain a large number of related challenging problems).

My approach in the Solutions section is often fairly heuristic, and I focus on the conceptual probabilistic reasoning. I think that — especially with natural science students — purely technical issues are in fact not usually the major obstacle: after all, the mathematical level of most elementary probability applications is not too advanced when compared to, say, even introductory fluid dynamics. Rather, the hardest part is often to develop an adequate understanding of, and intuition for, the characteristic patterns of reasoning and representing typically used in probability. Thus, technical tools such as derivatives or Taylor series are used freely whenever this seemed practical, but in a clearly informal, engineer-like, and nonrigorous way. My advice is: whenever you hit upon a problem or a solution that seems too technical, simply skip it and try the next one.

To specify the background required for this book, a fair number of problems collected here require little more than elementary probability and straight logical reasoning. Other problems clearly assume some familiarity with notions such as, say, generating functions, or the fact that around the origin $e^x \approx 1+x$. It is probably fair to say that anybody with a basic knowledge of probability, calculus, and statistics will benefit from trying to solve the problems posed here. Of course, you must try any way of attack that might promise success — including considering significant special cases and concrete examples — and resist the temptation to look up the solution! To help you along this way, I have provided a separate section of short hints, which indicate a direction into which you might orient your attention.

Potsdam, April 2007   *Wolf Schwarz*

Contents

3   Solutions 35

0

Notation and Terminology

P	probability.
rv	random variable, usually set in boldface, e.g., **T**, **N**.
DF	the distribution function of an rv, $F(t) = P(\mathbf{T} \leq t)$.
pdf	the probability density function of a continuous rv; if the DF, say F, of $\mathbf{T}$ is differentiable, then the pdf $f(t) = F'(t)$.
E	the expectation of an rv. For a continuous rv $\mathbf{T}$ with pdf f we have $\mathsf{E}[\mathbf{T}] = \int_{-\infty}^{\infty} t \cdot f(t)\, dt$, usually denoted as μ.
Var	the variance of an rv, $\mathsf{Var}[\mathbf{T}] = \mathsf{E}[(\mathbf{T} - \mathsf{E}[\mathbf{T}])^2]$, is usually denoted as σ^2.
i.i.d. rvs	random variables that are independently and identically distributed.
geometric rv	a positive discrete rv $\mathbf{N}$ with $P(\mathbf{N} = n) = p \cdot (1-p)^{n-1}$, where $n = 1, 2, \ldots$. Its mean is $1/p$, its variance $(1-p)/p^2$.
Poisson rv	a nonnegative discrete rv with $P(\mathbf{N} = n) = \lambda^n \cdot e^{-\lambda}/n!$, where $n = 0, 1, 2, \ldots$. Its mean and variance are $\lambda > 0$.
normal rv	a continuous rv with pdf

$$n(x|\mu, \sigma^2) = \frac{1}{\sigma\sqrt{2\pi}} \exp\left[-\frac{1}{2}\left(\frac{x-\mu}{\sigma}\right)^2\right]$$

	Its mean is μ, its variance σ^2. The case $\mu = 0, \sigma = 1$ defines a standard normal rv $\mathbf{Z}$, the DF of which is usually written as $P(\mathbf{Z} \leq z) = \Phi(z)$.
exponential rv	a nonnegative continuous rv with pdf $f(t|\lambda) = \lambda e^{-\lambda t}$. λ is called the *rate* of the distribution; an alternative notation replaces $1/\lambda$ with μ. Its mean is $1/\lambda$, its variance $1/\lambda^2$. The DF is $F(t|\lambda) = 1 - e^{-\lambda t}$.
gamma rv	the sum of n independent exponential rvs with rate λ. Its density is $f(t|\lambda, n) = \lambda e^{-\lambda t}\,(\lambda t)^{n-1}/(n-1)!$. Its mean is n/λ, its variance n/λ^2.

mgf   for a continuous rv $\mathbf{T} \geq 0$ the moment generating function $g(s)$, $s \geq 0$, is the Laplace transform of its pdf f, that is, $g(s) = \int_0^\infty e^{-st} f(t)\, dt$.

pgf   the probability generating function $g(z)$ of a nonnegative discrete rv $\mathbf{N}$ is the expectation of $z^{\mathbf{N}}$, that is, $g(z) = \mathsf{E}[z^{\mathbf{N}}] = \sum_{n=0}^\infty z^n \cdot \mathsf{P}(\mathbf{N} = n)$, $0 \leq z \leq 1$. In particular, $g'(1) = \mathsf{E}[\mathbf{N}]$.

tail formula   for a continuous rv $\mathbf{T} \geq 0$ with DF F, we have $\mathsf{E}[\mathbf{T}] = \int_0^\infty [1 - F(t)]\, dt$; also, $\mathsf{E}[\mathbf{T}^2] = 2 \int_0^\infty t\, [1 - F(t)]\, dt$.

ML estimate   the maximum likelihood (ML) estimate of a parameter; obtained by maximizing the likelihood of the data with respect to that parameter.

Bayes' theorem   states that for any two events A and B with $P(A) > 0$ we have $P(B|A) = P(A|B) \cdot P(B)/P(A)$.

hazard function   the hazard function $h(t)$ of a positive rv $\mathbf{T}$ with density f and DF F is defined as the conditional density at t, given that $\mathbf{T} \geq t$. That is, $h(t) = f(t)/[1 - F(t)]$.

1

Problems

1.1 To Begin or Not to Begin?

An urn contains k black balls and a single red ball. Peter and Paula draw without replacement balls from this urn, alternating after each draw until the red ball is drawn. The game is won by the player who happens to draw the single red ball. Peter is a gentleman and offers Paula the choice of whether she wants to start or not. Paula has a hunch that she might be better off if she starts; after all, she might succeed in the first draw. On the other hand, if her first draw yields a black ball, then Peter's chances to draw the red ball in his first draw are increased, because then one black ball is already removed from the urn. How should Paula decide in order to maximize her probability of winning?

1.2 A Tournament Problem

Ten players participate in the first round of a tennis tournament: 2 females and 8 males. Five single matches are fixed at random by successively drawing, without replacement, the names of all 10 players from an urn: the player drawn first plays against the one whose name comes up second, the third against the fourth, etc.

a. What is the probability that there will not be a single match involving two female players? Is this probability smaller, equal to, or larger than the corresponding probability with 20 females and 80 males?

b. Try to answer the general case in which there are $2n$ players, of whom $2 \leq k \leq n$ are female. What is the probability $p(k, n)$ that among the n matches there will not be a single one involving two female players?

1.3 Mean Waiting Time for $1 - 1$ vs. $1 - 2$

Peter and Paula play a simple game of dice, as follows. Peter keeps throwing the (unbiased) die until he obtains the sequence $1 - 1$ in two successive throws. For Paula, the rules are similar, but she throws the die until she obtains the sequence $1 - 2$ in two successive throws.

a. On average, will both have to throw the die the same number of times? If not, whose expected waiting time is shorter (no explicit calculations are required)?

b. Derive the actual expected waiting times for Peter and Paula.

1.4 How to Divide up Gains in Interrupted Games

Peter and Paula play a game of chance that consists of several rounds. Each individual round is won, with equal probabilities of $\frac{1}{2}$, by either Peter or Paula; the winner then receives one point. Successive rounds are independent. Each has staked $50 for a total of $100, and they agree that the game ends as soon as one of them has won a total of 5 points; this player then receives the $100. After they have completed four rounds, of which Peter has won three and Paula only one, a fire breaks out so that they cannot continue their game.

a. How should the $100 be divided between Peter and Paula?

b. How should the $100 be divided in the general case, when Peter needs to win a more rounds and Paula needs to win b more rounds?

1.5 How Often Do Head and Tail Occur Equally Often?

According to many people's intuition, when two events, such as head and tail in coin tossing, are equally likely then the probability that these events will occur equally often increases with the number of trials. This expectation reflects the intuitive notion that in the long run, asymmetries of the frequencies of head and tail will "balance out" and cancel.

To find the basis of this intuition, consider that $2n$ fair and independent coins are thrown at a time.

a. What is the probability of an even $n : n$ split for head and tail when $2n = 20$?

b. Consider the same question for $2n = 200$ and $2n = 2000$.

1.6 Sample Size vs. Signal Strength

An urn contains six balls — three red and three blue. One of these balls — let us call it ball A — is selected at random and permanently removed from the urn without the color of this ball being shown to an observer. This observer may now draw successively — at random and with replacement — a number of individual balls (one at a time) from among the five remaining balls, so as to form a noisy impression about the ratio of red vs. blue balls that remained in the urn after A was removed.

Peter may draw a ball six times, and each time the ball he draws turns out to be red. Paula may draw a ball 600 times; 303 times she draws a red ball, and 297 times a blue ball. Clearly, both will tend to predict that ball A was probably blue. Which of them — if either — has the stronger empirical evidence for his/her prediction?

1.7 Birthday Holidays

The following problem is described in Cacoullos (1989, pp. 35–36).

A worker's legal code specifies as a holiday any day during which at least one worker in a certain factory has a birthday. All other days are working days. How many workers (n) must the factory employ so that the expected number of working man-days is maximized during the year?

1.8 Random Areas

Peter and Paula both want to cut out a rectangular piece of paper. Because they are both probabilists they determine the exact form of the rectangle by using realizations of a positive rv, say U, as follows. Peter is lazy and generates just a single realization of this rv; he then cuts out a square that has length and width equal to this value. Paula likes diversity and generates two independent realizations of U. She then cuts out a rectangle with width equal to the first realization and length equal to the second realization.

    a. Will the areas cut out by Peter and Paula differ in expectation?

    b. If they do, is Peter's or Paula's rectangle expected to be larger?

1.9 Maximize Your Gain

A nonnegative rv U has DF F and density $f = F'$; its mean μ and variance σ^2 are both finite. A game is offered, as follows: you may choose a nonnegative number c; if $U > c$ then you win the amount c, otherwise you win nothing.

As an example, suppose U is the height (measured in cm) of the next person entering a specific public train station. If you choose $c = 100$ then you will almost surely win that amount. A value of $c = 200$ would double your amount if you win, but of course drastically reduce your winning probability.

a. Find an equation to characterize the value of c that maximizes the expected gain.

b. Give a characterization of the optimal value of c in terms of the hazard function of U (see page 2 for the definition of the hazard function).

c. Derive c explicitly for an exponential rv with rate λ (see page 1 for a definition). How large is the maximum expected gain?

1.10 Maximize Your Gain When Losses Are Possible

Under the same assumptions as in Problem 1.9, the rule of the game is changed, as follows: if $U > c$ an amount of c is won, but otherwise the amount c is lost.

a. Characterize the value of c that maximizes the expected gain, $G(c)$.

b. Can the game be unfavorable, i.e., can (even) the maximum expected gain become negative?

c. How large is the maximum expected gain for an exponential rv with rate λ? Compare this to the solution of the previous problem; explain the difference.

1.11 The Optimal Level of Supply

A man offers milk to the spectators of the weekly baseball matches. Before each match, he orders c units of milk at a price of $b > 0$ per unit; during the match he sells each unit for a price of $s > b$. Because the milk cannot be kept fresh for a week, each unsold unit imposes a loss of b. The actual demand D of milk varies from week to week, according to a strictly increasing DF F with density f. Assume that $F(0) = 0$ and that D is a continuous rv with finite mean.

a. For given prices b, s, and a given demand structure F, the milkman can choose among different values of c. Consider c as a continuous variable, and denote as $G(c)$ the expected net gain as a function of c. Discuss the shape and qualitative properties of G.

   b. What is the optimal amount of milk, c_{opt}, the man should stock in order to maximize his expected net gain?

1.12 Mixing RVs vs. Mixing Their Distributions

   The concept of a "mixture distribution" is used in probability and its applications in at least two different ways that have quite different meanings.

   Let $\mathbf{X}$ and $\mathbf{Y}$ be two independent normal rvs, with means $\mu_x = 50$ and $\mu_y = 150$ and standard deviations of $\sigma_x = \sigma_y = 10$. Consider the rv $\mathbf{A}$ defined by

$$A = \frac{1}{2}(X + Y)$$

The idea here is that in each realization both a value of $\mathbf{X}$ and a value of $\mathbf{Y}$ are generated, and half of each is added to produce the value of $\mathbf{A}$. Thus, in a fairly direct sense each individual realization of $\mathbf{A}$ contains (is made up of) one part of $\mathbf{X}$ and one part of $\mathbf{Y}$ — in this sense, $\mathbf{A}$ is a 50/50 mixture of $\mathbf{X}$ and $\mathbf{Y}$, much like one mixes half a pound of butter and half a pound of flour.

   Next, consider an rv $\mathbf{B}$ that comes, in each realization, with probability $\frac{1}{2}$ from a normal distribution (namely, that of $\mathbf{X}$) with mean $\mu = 50$ and standard deviation 10 and with probability $\frac{1}{2}$ from a normal distribution (namely, that of $\mathbf{Y}$) with mean $\mu = 150$ and standard deviation 10. Thus, its density is equal to

$$f_B(b) = \frac{1}{2}\left[n(b|\mu = 50, \sigma = 10) + n(b|\mu = 150, \sigma = 10)\right]$$

where $n(\cdot|\mu, \sigma)$ is the normal density with mean μ and standard deviation σ. The idea here is that in each realization either a value of $\mathbf{X}$ or a value of $\mathbf{Y}$ is generated (but not of both) with equal probability, and that this value then determines that of $\mathbf{B}$. However, across many realizations, the density of $\mathbf{B}$ will still represent a 50/50 mixture of $\mathbf{X}$ and $\mathbf{Y}$.

   a. Sketch the densities of $\mathbf{A}$ and $\mathbf{B}$. Is $\mathbf{A}$ normally distributed? Is $\mathbf{B}$?

   b. Derive the means and variances of $\mathbf{A}$ and $\mathbf{B}$. Compare and explain.

   c. Let $\mathbf{X}$, $\mathbf{Y}$ be two arbitrary but independent rvs with densities f_X, f_Y; means μ_x, μ_y; and standard deviations σ_x, σ_y. Let $0 \le p \le 1$ be a proportion mixing either the rvs themselves:

$$A = p \cdot X + (1 - p) \cdot Y$$

or their densities

$$f_B(b) = p \cdot f_X(b) + (1 - p) \cdot f_Y(b)$$

Derive for this more general case the means and variances of $\mathbf{A}$ and $\mathbf{B}$.

1.13 Throwing the Same vs. Different Dice

The standard binomial sampling scheme assumes n independent trials with a constant probability of success. Suppose, for example, that we are given a single fair die, and that the success event consists in throwing, say, a 5. Thus, each single throw results in a success with probability $p = \frac{1}{6}$. Within $n = 144$ independent trials, we would thus expect about $n \cdot p = 24$ successes, and the variance of the number of successes would be equal to $n \cdot p \cdot (1 - p) = 20$.

a. Suppose we are given two biased dice. Die A will show a 5 with probability $p = \frac{1}{4}$, and die B shows a 5 with probability $p = \frac{1}{12}$. With these two dice 144 trials are conducted, as follows. For the first 72 trials we use die A, and during the last 72 trials we use die B. Given that the average success probability across both dice is equal to $(\frac{1}{4} + \frac{1}{12})/2 = \frac{1}{6}$, we would again expect 24 successes. Is the variance of the number of successes larger than, equal to, or smaller than with the standard binomial sampling scheme?

b. Again we are given the two dice A and B described in a. This time, however, we select at random one die and then throw this selected die 144 times, observing again the number of throws yielding a 5. Is the variance of the number of successes larger than, equal to, or smaller than with the standard binomial sampling scheme?

c. Once again we are given the two dice A and B. This time, however, in each of 144 trials we start by choosing at random one of the two dice, then throw it and observe whether or not the trial yields a 5. Is the variance of the number of successes larger than, equal to, or smaller than with the standard binomial sampling scheme?

1.14 Random Ranks

Peter draws $n = 100$ independent realizations of a continuous rv and ranks them in increasing order from 1 to 100. Subsequently, Paula draws a single value from the same population and inserts this value into the rank order created earlier by Peter. For example, if her value is such that 50 of Peter's draws are smaller and 50 are larger, then the rank associated with her draw would be 51 — that is, overall, her value would be the 51st in increasing order. Or, if her value is smaller than all 100 of Peter's, then the rank 1 would be associated with it.

a. Is it more likely that Paula's value will occupy rank 51 than rank 1?

b. Derive for general n the probability that Paula's value will occupy rank k, where $1 \leq k \leq n + 1$.

1.15 Ups and Downs

Ups and downs occur in probability just as in real life. An elementary probability version with the real-life property that ups are (more) often followed by downs is as follows.

Let $(\mathbf{A}, \mathbf{B}, \mathbf{C})$ be three independent and identically distributed continuous rvs that are realized sequentially: first $\mathbf{A}$, then $\mathbf{B}$, and finally $\mathbf{C}$. Let us say that an *increment* occurs with $\mathbf{B}$ if $\mathbf{B} > \mathbf{A}$, and a decrement otherwise. Similarly, an increment occurs with $\mathbf{C}$ if $\mathbf{C} > \mathbf{B}$, and a decrement otherwise.

a. Suppose you are told only that $\mathbf{B}$ has led to an increment, but not the actual value of $\mathbf{B}$. Argue that conditional on this information $\mathbf{C}$ is twice as likely to yield a decrement than an increment — even though $\mathbf{B}$ and $\mathbf{C}$ are independent.

b. Random variables are said to be *exchangeable* if their joint density is the same for any permutation of its arguments. Thus, if $(\mathbf{A}, \mathbf{B}, \mathbf{C})$ are exchangeable, then their density, say $f(a, b, c)$, is the same for any permutation of (a, b, c). Note that exchangeable rvs may still be dependent. Does the property described in a. still hold if $(\mathbf{A}, \mathbf{B}, \mathbf{C})$ are exchangeable?

c. Suppose that the three rvs have the same marginal distribution with mean μ and variance σ^2, and let them have the common pairwise correlation ϱ. Show that the correlation of the rvs $\mathbf{B} - \mathbf{A}$ and $\mathbf{C} - \mathbf{B}$ is generally equal to $-\frac{1}{2}$.

1.16 Is 2X the Same as $\mathbf{X_1} + \mathbf{X_2}$?

Let $\mathbf{X}$ be a continuous rv with density f, and let $\mathbf{X_1}$, $\mathbf{X_2}$ be two independent rvs, both distributed as is $\mathbf{X}$. It is then not usually the case that the rv $2\mathbf{X}$ is distributed as is $\mathbf{X_1} + \mathbf{X_2}$. However, the Cauchy density whose standardized form is given by

$$f(x) = \frac{1}{\pi} \cdot \frac{1}{1 + x^2}$$

possesses this property: $\mathbf{X_1} + \mathbf{X_2}$ has the same distribution as the rv $2\mathbf{X}$. It is illustrated and compared to the standard normal distribution in Figure 1.1.

a. Based on the property described earlier, argue without any explicit calculation that the variance of the Cauchy distribution is necessarily infinite.

b. Give an inductive argument for the rather unintuitive feature that for the Cauchy distribution the arithmetic mean from a sample of $n = 2^k$ independent realizations of $\mathbf{X}$ has exactly the same distribution as each contributing summand itself.

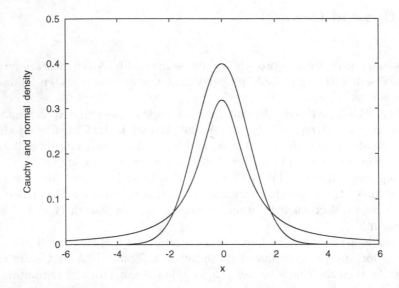

Fig. 1.1. The standard Cauchy and the standard normal density look basically similar: they are both unimodal and symmetric around $x = 0$. However, the Cauchy has thicker tails, whereas the normal density is more concentrated around its mean.

1.17 How Many Donors Needed?

To organize a charity event that costs $100, an organization raises funds. Independent of each other, one donor after another gives some amount of money (considered as a continuous quantity here) that is exponentially distributed (for definition, see page 1), with a mean of $20. The process is stopped as soon as $100 or more has been collected. Find the distribution, mean, and variance of the number of successive donors needed until at least $100 dollars has been collected.

1.18 Large Gaps

Denote as $\{N(t), t \geq 0\}$ a Poisson process with rate λ. Let $\mathbf{W}$ be the waiting time until for the first time no event in $\{N(t)\}$ has occurred for the last τ time units. Derive $\mathsf{E}[\mathbf{W}]$. For example, $\{N(t)\}$ might stand for the sequence of cars passing at a crosswalk. A pedestrian needs τ seconds to cross the road — how long does it on average take before he has reached the other side of the road?

1.19 Small Gaps

Related to the preceding problem, let **U** be the waiting time until two events occur within τ time units. Derive $\mathsf{E}[\mathbf{U}]$.

In some applications, the event that has latency **U** is called a "coincidence." For example, a volume of biological tissue could be permanently destroyed when two damaging particles are absorbed within τ (or less) time units. The idea here is that following the first absorption the tissue needs to recover for τ time units; this opens a window of vulnerability during which a further (second) particle has a lethal effect.

1.20 Random Powers of Random Variables

Let **M, N** be independent Poisson rvs (for definition, see page 1), with parameters (means) α, $\beta \geq 0$, respectively. Define a rv $\mathbf{Q} \equiv \mathbf{M}^{\mathbf{N}}$.

a. Find the expectation $\mathsf{E}[\mathbf{Q}]$.

b. Does $\mathsf{E}[\mathbf{Q}]$ generally increase as a function of α and β? Explain.

1.21 How Many Bugs Are Left?

Peter and Paula are given copies of the same text for independent proofreading. Peter finds 20 errors, and Paula finds 15 errors, of which 10 were found by Peter as well. Estimate the number of errors remaining in the text that have not been detected by either Peter or Paula.

1.22 ML Estimation with the Geometric Distribution

Let the discrete rv **K** have a geometric distribution (see page 1),

$$\mathsf{P}(\mathbf{K} = k) = p \cdot (1 - p)^{k-1} \quad , \quad k = 1, 2, \ldots$$

and let $\{k_i; i = 1, \ldots, n\}$ be n independent realizations from it.

a. Find the maximum likelihood (ML) estimate (for definition, see page 2), $\hat{p}$, of p. Give a simple argument why $\hat{p}$ overestimates p.

b. Determine the asymptotic standard error, $a.s.e.(\hat{p})$, of this estimate as a function of p. Consider the asymptotic standard error as $p \to 0$ and as $p \to 1$. For which value of p is $a.s.e.(\hat{p})$ maximal?

c. Explain the apparent paradox that $\hat{p}$ is, on the one hand, the reciprocal of a sum of rvs (which in turn will tend to normality), and on the other hand that it is itself asymptotically normally distributed.

1.23 How Many Twins Are Homozygotic?

Twin pairs are either homozygotic (with probability α) or dizygotic (probability $1 - \alpha$). In the first case, their common sex is genetically determined only once, for both twins together, so that they must necessarily be of the same sex. In the second case, the twins' sexes are determined independently and therefore could potentially be different. Suppose that generally in any genetic sex determination in a given population the outcome will be "male" (m) with probability β and "female" (f) with probability $1 - \beta$.

Obviously, the sex-related classification of any given twin pair (i.e., mm vs. ff vs. mf) is readily apparent, whereas the determination of their homo- vs. dizygotic status requires elaborate genetic diagnostics. Specifically, if the two twins are of opposite sexes (i.e., the pair is mf) then they must necessarily be dizygotic, but if they are not (i.e., an mm or ff pair) then they may be either homo- or dizygotic.

a. L. v. Bortkiewicz (1920; for a summary, see von Mises, 1931, p. 407) has collected the sex-related classification for a total of $n = 17798$ twin pairs born in Berlin from 1879 to 1911. The frequencies were $n(mm) = 5844$, $n(ff) = 5612$, and $n(mf) = 6342$. From these data, how could one estimate the latent probability α of a twin pair to be homozygotic?

b. Is the estimate of β significantly different from $\frac{1}{2}$?

1.24 The Lady Tasting Tea

In 1935, R.A. Fisher presented the famous "lady tasting tea" problem[1]. A lady claims she can tell whether in a tea-plus-milk infusion the tea (type T cup) or the milk (type M cup) was poured in first. To test this claim, the lady is presented with n cups of type T and n cups of type M; her task is to tell these two sets — which differ in nothing other than the order in which tea and milk were poured in — apart. More specifically, after she has been informed that there is an equal number of T and M cups, she first tastes the content of each of the $2n$ cups and then singles out those n of them that she thinks are, say, of type T; this implies that she considers the remaining n cups to be of type M.

As a simple model of this situation, suppose the lady can identify the true state of any given cup with probability θ, $0 \leq \theta \leq 1$; with probability $1-\theta$, she can only guess. Of course, the skeptic's claim would be that $\theta = 0$. Judgments about successive cups are considered to be independent.

a. Define a "hit" to occur when a cup is correctly classified. Explain why the number of hits is necessarily the same for both the set of n cups the lady

---

[1] FISHER, R.A. (1935).

designates as type T and the remaining set of n cups she considers to be of type M. Therefore, if we know the number of hits in the first set, the number of hits in the second set is redundant.

b. What is the probability $p_n(k|\theta)$ that among the n cups she selects as being of type T, there will be exactly k, $0 \le k \le n$, hits, i.e., cups that are in fact of type T?

c. Suppose the lady's ability to tell T and M apart corresponds to $\theta = 0.50$ in the preceding model. Thus, while not perfect, she clearly does much better than guessing. Using the largest possible significance level below the conventional $\alpha = 0.05$, how large is her probability $\beta(n)$ to demonstrate her ability in an experiment involving $n = 4, 10$, and 20 cups of each type? (A computer is required to do these calculations.)

1.25 How to Aggregate Significance Levels

Suppose a one-sided statistical test is based on a statistic Λ that under the H_0 has the distribution function F, the rejection region being formed by large values of Λ. The test is applied to a first set of data and the statistic $\Lambda = \lambda_1$ is observed, with an associated observed p-value of $p_1 = P(\Lambda > \lambda_1) = 0.09$. On a second independent set of data the same test is applied, and this time a statistic $\Lambda = \lambda_2$ is observed, with an associated observed p-value of $p_2 = P(\Lambda > \lambda_2) = 0.07$.

a. Argue that the observed p-value is a rv that is uniformly distributed on $[0, 1]$ if H_0 is true.

b. From a., derive a way to aggregate p_1, p_2 into a single overall p-value.

c. With the procedure from b., describe the set of $\{(p_1, p_2)\}$ that is judged overall-significant at some given theoretical significance level α, such as $\alpha = 0.05$.

1.26 Approximately How Tall Is the Tallest?

Let $\mathbf{U}$ be a continuous rv with strictly increasing DF F, and let $\mathbf{M}_n$ be the largest of n independent realizations of $\mathbf{U}$.

a. Determine the probability of the event $\{\mathbf{M}_n \le F^{-1}(1 - x/n)\}$, where F^{-1} is the inverse function of F.

b. From a., derive an approximation, valid for medium and large n, for the DF, say $G_n(m)$, of $\mathbf{M}_n$.

c. Show that if $\mathbf{U}$ is exponentially distributed, then $\mathbf{M}_n$ tends to the double exponential distribution.

d. Suppose the height (measured in cm) of males in a country is normally distributed, with mean $\mu = 176$ cm and standard deviation $\sigma = 8$ cm. Sketch the approximate DF of the height of the tallest male, if there are (i) 4 million males (as in a small country) or (ii) 120 million males (as in a large country). Compare these graphs to the corresponding figures for females, assuming that their height is normally distributed, with mean $\mu = 170$ cm and standard deviation $\sigma = 7$ cm.

1.27 The Range in Samples of Exponential RVs

Let $\mathbf{T}$ be an exponential rv with density $\lambda e^{-\lambda t}$; the parameter λ is called the *rate* of the density (cf. page 1). It is not difficult to show that $\mathsf{E}[\mathbf{T}] = 1/\lambda$. Exponential rvs have two interesting properties.

(i) The minimum of independent exponential rvs, which may have different rates, is again exponentially distributed, with a rate equal to the sum of the individual rates. For example, the minimum of two independent exponential rvs with rates λ_1 and λ_2 is again an exponential rv, with rate $\lambda_1 + \lambda_2$.

(ii) If an exponential rv $\mathbf{T}$ is larger than some other positive independent rv — let us call it $\mathbf{Q}$ — then the *excess* $\mathbf{T} - \mathbf{Q}$ is again exponentially distributed with rate λ. This characteristic feature holds also, as a special case, when $\mathbf{Q}$ is a constant, say q. It is sometimes called the "lack-of-memory" property: given that $\mathbf{T}$ exceeds $\mathbf{Q}$, then its *additional* lifetime, from there on, has exactly the same distribution as the original lifetime — just as if the process had "no memory" for the already-elapsed period of time.

We draw a sample of n i.i.d. realizations of $\mathbf{T}$ (for the meaning of i.i.d., see page 1). Let the rv $\mathbf{T}_{(i)}$, $i = 1, \ldots, n$ be the ith smallest value in this sample. The positive rv $\mathbf{R}_n = \mathbf{T}_{(n)} - \mathbf{T}_{(1)}$ is the largest minus the smallest value; it is called the *range* of the sample. Of course, the value of $\mathbf{R}_n$ will vary from sample to sample, according to some distribution that depends on λ and, of course, on n.

a. Use the two properties described earlier heuristically to derive an expression for the expectation of $\mathbf{R}_n$.

b. Similarly, reason heuristically to find the distribution function of $\mathbf{R}_n$ without explicit calculations.

1.28 The Median in Samples of Exponential RVs

As in Problem 1.27, let $\mathbf{T}$ be an exponential rv with density $\lambda e^{-\lambda t}$. We draw a sample of size $n = 2k - 1$ (k a positive integer) i.i.d. realizations of $\mathbf{T}$, and estimate the median of $\mathbf{T}$ by the kth order statistic of this sample, i.e., the midmost element of the sample that is both larger and smaller than $k - 1$ other realizations. Call this median estimate $\mathbf{M}$.

a. Derive the density of **M**.

b. Find an expression for the expectation of **M**, and show that for small k it is severely biased.

1.29 Breaking the Record

Consider a process in which, in a chronological (sequential) order, the i.i.d. realizations of an rv **X** with parent density f and DF F are generated, yielding the sequence $\{\mathbf{X}_i; i = 1, 2, \ldots\}$. From time to time, it will then be the case that a realization $\mathbf{X}_i$ occurs that is larger than the largest value that has been seen so far. In accordance with the everyday meaning of this term, such a realization is called a *record*. By this definition, $\mathbf{X}_1$ is necessarily a record. Also, the second record is the first realization of **X** that is larger than $\mathbf{X}_1$ was. In general, the rth record is the first realization of **X** that is larger than the $(r-1)$th record. Denote the density of the rth record as $g_r(x)$.

a. Consider a sequence of $n = 100$ realizations of **X**. How many records could there be minimally and maximally? Give a simple recursive argument, relating the cases $n-1$ and n, to find how many records one would on average expect to see in $n = 100$ realizations. Explain why the number of records is independent of the parent distribution f.

b. Derive (in terms of f, F) the density of the second record, $g_2(x)$.

c. Give an inductive argument to establish the general result that the density

$$g_r(x) = f(x) \cdot \frac{1}{(r-1)!} \cdot \{-\ln[1 - F(x)]\}^{r-1}$$

d. Explain that in the exponential case of $f(x) = \lambda e^{-\lambda x}$ the result derived in c. could be anticipated without explicit calculations from the lack-of-memory property described in the statement of Problem 1.27.

1.30 Paradoxical Contribution

Two large predatory birds, A and B, feed on the same habitat. Bird A's daily prey (in grams) is normally distributed with mean $\mu_A = 60$ g and $\sigma_A = 5$; for bird B, $\mu_B = 40$ g and $\sigma_B = 10$. Thus, on average, bird A is a more successful predator; also, its amount of daily prey is less variable than that of bird B. Clearly, on average, the daily overall prey of both birds together equals 100 g. However, as both birds feed on the same habitat, the correlation, ϱ, of their daily prey is negative across days, say $\varrho = -0.8$: any given animal in the habitat that has fallen prey to A is no longer available to

B, and vice versa. Let us assume that the situation is adequately described by a bivariate normal distribution.

On an unusually successful day, A and B together manage to reap 175 g.

a. Try to estimate from the information given how much, in expectation, bird A's share was on days in which A and B together have reaped 175 g.

b. Try to answer the question in a. if the birds feed on different habitats so that $\varrho = 0$.

1.31 Attracting Mediocrity

Peter has an IQ of 90 whereas the IQ of Paula is 110. However, due to unsystematic biological or psychological day-to-day variation that is unrelated to the IQ per se, any single measurement of either IQ is distorted by an independent additive measurement error that has a zero-mean normal distribution with variance σ^2. For example, if Paula's IQ were measured repeatedly, the outcomes would be normally distributed with a mean of 110 (her "true" IQ) and a standard deviation of σ.

a. Suppose that either Peter or Paula is selected at random ($p = \frac{1}{2}$), and his/her IQ is measured. You do not know who was selected, but you are told that the result of this first measurement is 105. Now the *same* person — whose identity is unknown to you — is measured a second time. What is your prediction for the outcome of this second measurement if $\sigma = 3$?

b. Answer the same question if $\sigma = 20$.

c. Suppose that instead of just contemplating Peter and Paula we now deal with a large population of individuals whose true IQs (an rv that we may call **T**) is normally distributed with mean μ_T and variance σ_T^2. As before, each individual measurement is distorted by an independent and additive normally distributed error **E** that has zero-mean and variance σ^2. A single person is drawn at random from this large population, his/her IQ is measured, and the outcome is some above-average value $x > \mu_T$. Show that when the IQ of that same person is measured a second time, the expected outcome of the second ("repeated") measurement is larger than μ_T but smaller than x.

1.32 Discrete Variables with Continuous Error

Sometimes an underlying latent (i.e., not directly observable) rv **N** of theoretical or practical interest is generically integer-valued, such as a binomial, Poisson, or geometrical rv. However, measurements **S** of it are distorted by a random error **E** that is continuous. For example, a random integer number of coins of one type (e.g., dimes) is inserted into a vending machine with a control balance that registers the total weight, say in units of the nominal weight of a single dime. In this example, there will be an integer number of coins, but due to minting imperfections, wear-off, and soiling, each coin will differ somewhat from its ideal nominal weight. Summing the individual deviations from the nominal weight across all coins we get the overall measurement error **E**.

If we assume that the measurement error **E** (thought to be independent of **N**) has a normal distribution with zero mean and standard deviation $\sigma > 0$, then the actually observed measurements are in fact realizations of the rv **S** = **N** + **E**. Let f be the density of **S**.

a. Characterize in words or with a figure the changes of the density f as σ varies from a near-zero value to larger values. When, approximately, will f change from multi- to unimodality? To fix ideas, consider the example of a geometric distribution, $P(\mathbf{N} = n) = p \cdot (1 - p)^{n-1}$, $p = 0.2$.

b. Argue that f can also be interpreted as a discrete mixture of normal densities with means $\mu = n$ and common standard deviation σ, the mixture weights being given by $P(\mathbf{N} = n)$.

1.33 The High-Resolution and the Black-White View

Ideally, many properties and processes of the real world show extremely fine gradations and variations. However, in order to measure and to process this information, we are often forced to simplify and reduce it. For example, many measurement devices digitize quantities (such as force, voltage, or time) that in theory are conceptualized as continuous variables. An extreme form of data compression is *thresholding*, which reduces a continuous input into a simple binary output. Two important questions then are: how should this threshold be set, and how much information do we lose along the way? The following problem deals with these two questions in a simple but exemplary context.

Consider a sample of n independent realizations $\{t_i, i = 1, \ldots, n\}$ of an exponential rv with density $f(t|\mu) = (1/\mu) \exp(-t/\mu)$. Researchers A and B both seek to estimate the unknown mean, μ. A observes the n original raw data and can make use of the full information to estimate μ. B on the other hand obtains the data only after they have passed a digital filter with a threshold-type mechanism, such that each realization is classified as 0 if it is $\leq c$, and classified as 1 if it is $> c$. Suppose k, $0 < k < n$, of the n realizations turned out to be $\leq c$.

a. Derive the ML estimate of μ and its asymptotic standard error (a.s.e.) from the raw data as seen by researcher A.

b. For a given threshold value c, derive the ML estimate of μ from the digitized data as seen by researcher B. Does this estimate correspond to your intuition?

c. Find the a.s.e. of this estimate from the data as seen by researcher B.

d. For a given value of μ what is the "best" threshold value of c, i.e., the one that minimizes the a.s.e. determined in part c.? How large, relative to the estimate in a., is this a.s.e.?

1.34 The Bivariate Lognormal

The standard parametric model for simple correlation and regression contexts is the bivariate normal distribution. It is quite interesting to see how important characteristics regarding, e.g., the correlation coefficient change under a different bivariate distribution model, such as the lognormal distribution.

Let $\mathbf{X}$ be a normal rv with mean μ_x and standard deviation σ_x. Then the rv $\mathbf{U} = \exp(\mathbf{X})$ has a univariate lognormal distribution. A basic result about the rv $\mathbf{U}$ is that its expectation is equal to $\mathsf{E}[\mathbf{U}] = \exp(\mu_x + \frac{1}{2}\sigma_x^2)$.

Similarly, let the rvs $< \mathbf{X}, \mathbf{Y} >$ have the bivariate normal distribution with parameters $< \mu_x, \mu_y, \sigma_x, \sigma_y, \varrho >$, where μ_y, σ_y are the mean and standard deviation of $\mathbf{Y}$, and ϱ is the correlation of $\mathbf{X}$ and $\mathbf{Y}$. Then the pair $< \mathbf{U} = \exp(\mathbf{X}), \mathbf{V} = \exp(\mathbf{Y}) >$ has a bivariate lognormal distribution. Note that $< \mathbf{U}, \mathbf{V} >$ are a pair of nonnegative rvs.

a. Based on the result given earlier concerning the expectation of $\mathbf{U} = \exp(\mathbf{X})$, derive without new explicit calculations the variance of $\mathbf{U}$.

b. Assume that $< \mathbf{X}, \mathbf{Y} >$ are uncorrelated, $\varrho = 0$. In this case, are the rvs $< \mathbf{U}, \mathbf{V} >$ uncorrelated, too?

c. Explain why for $\varrho = +1$ the $< \mathbf{U}, \mathbf{V} >$ are perfectly correlated, too, if $\sigma_x = \sigma_y$. Why does this result not extend to $\varrho = -1$ as well, even though in the bivariate normal model the cases $\varrho = +1$ and $\varrho = -1$ are perfectly symmetric?

d. Using the results from a. and b., show that the correlation of $\mathbf{U}$ and $\mathbf{V}$ equals

$$\mathsf{corr}(\mathbf{U}, \mathbf{V}) = \frac{\exp(\varrho\sigma_x\sigma_y) - 1}{\sqrt{\left[\exp(\sigma_x^2) - 1\right]\left[\exp(\sigma_y^2) - 1\right]}}$$

1.35 The $\arcsin(\sqrt{p})$ Transform

Let the rv $\mathbf{N}$ be the number of successes in n independent trials, each with success probability p. Clearly, $\mathbf{N}$ is binomially distributed with parameters n and p, and $\hat{p} = \mathbf{N}/n$ is the usual estimate of p. It is unbiased — that is, $\mathsf{E}[\hat{p}] = p$ — and its variance is $\mathsf{Var}[\hat{p}] = p(1-p)/n$, which varies as a quadratic function of p. However, many applications — especially in analysis of variance and regression contexts — require variables that may differ in mean but not in variance across conditions. To circumvent this problem, it is customary to analyze the strictly increasing transformation $\arcsin(\sqrt{\hat{p}})$ rather than $\hat{p}$ itself.

a. Let $g(p)$ be a differentiable function of p. Argue heuristically that if n is large, then

$$g(\hat{p}) \; = \; g(p + \Delta p) \; \approx \; g(p) + \Delta p \cdot g'(p)$$

where Δp is the sampling error $\hat{p} - p$ of the estimate $\hat{p}$. From this representation, derive the approximate mean and variance of the rv $g(\hat{p})$.

b. At first glance, the choice of the transform $p \mapsto g(p) = \arcsin(\sqrt{p})$ seems rather exotic. Give a principled rationale for this particular choice.

1.36 Binomial Trials Depending on a Latent Variable

In many situations, the outcome of binomial trials is modeled through an underlying *latent* — that is, not directly observable — rv $\mathbf{X}$ with strictly increasing DF Φ, such that each of the binomial trials yields a success if and only if $\mathbf{X} \leq c$. For example, in a population of individuals the susceptibility to flu may follow a particular population distribution. A flu is successfully avoided by an individual i if his or her (standardized) susceptibility $\mathbf{X}_i$ does not exceed some critical, but unknown, flu threshold c. From observations of the relative flu frequency in a random sample taken from this population, one would then like to estimate the threshold parameter, c.

Suppose n binomial trials yield k successes, leading to the usual estimate $\hat{p} = k/n$ for the true success probability $p = \Phi(c)$. Thus, on equating $\hat{p} = \Phi(\hat{c})$, the natural estimate, $\hat{c}$, of the model parameter c is $\Phi^{-1}(\hat{p})$, where Φ^{-1} is the inverse function of Φ.

Of course, from sample to sample our estimate $\hat{c}$ will vary, just as the relative success probability, $\hat{p}$, of which it is a function. Determine the approximate standard error (i.e., the standard deviation) of the estimate $\hat{c}$ based on a sample of n trials if Φ is

a. the standard logistic DF, $\Phi(x) = 1/[1 + \exp(-x)]$,

b. the standard normal DF.

1.37 The Delta Technique with One Variable

Let $\mathbf{X}$ be an rv with expectation μ and variance σ^2, and let $f(\mathbf{X})$ be a given, known function of $\mathbf{X}$. In general then

$$\mathsf{E}[f(\mathbf{X})] \neq f[\mathsf{E}(\mathbf{X})]$$

a. Find a series-based approximation to $\mathsf{E}[f(\mathbf{X})]$ when σ is small so that $\mathbf{X}$ is concentrated in the neighborhood of μ.

b. Apply this result to the case of $f(x) = e^x$ when $\mathbf{X}$ is a normal rv with mean μ and variance σ^2. Compare this approximation to the exact result for this case, $\mathsf{E}[\exp(\mathbf{X})] = \exp(\mu + \frac{1}{2}\sigma^2)$, as given in Problem 1.34. When will the approximation be acceptable?

c. A machine is constructed to throw a ball vertically with an initial velocity of 120 [m/s]. However, due to imperfections of the machine, the actual angle varies from throw to throw according to a normal distribution with a mean of 90° and a standard deviation of 10°. What is, approximately, the expected maximum height that the ball will reach, neglecting the air resistance?

1.38 The Delta Technique with Two Variables

Problem 1.37 can be generalized to functions of more than one rv. Thus, let $\mathbf{X}$, $\mathbf{Y}$ be two rvs with expectations μ_x, μ_y, variances σ_x^2, σ_y^2, and correlation ϱ. Also, let $f(\mathbf{X}, \mathbf{Y})$ be a given function of these rvs.

a. Find a series-based approximation to $\mathsf{E}[f(\mathbf{X}, \mathbf{Y})]$ that holds when $\mathbf{X}$ and $\mathbf{Y}$ are concentrated in a region around the point (μ_x, μ_y).

b. Apply this result to the case of the ratio of two independent rvs, i.e., $f(x, y) = x/y$ and $\varrho = 0$.

c. Explain how the solution in part b. may also be obtained from the solution of Problem 1.37, related to functions of a single rv.

d. Consider the special case of an rv defined as

$$\mathbf{F}_{m,n} = \frac{\mathbf{U}/m}{\mathbf{V}/n}$$

where $\mathbf{U}$, $\mathbf{V}$ are independent χ^2-rvs, with m, n degrees of freedom, respectively. Compare for this case the approximation to the exact result, namely the mean of the F-distribution, which is $n/(n-2)$.

e. Suppose that in question c. of problem Problem 1.37 in addition to, and independent of, the variation of the angle α, the initial velocity v_0 of the ball varies according to a normal distribution with $\mu_{v_0} = 120$ [m/s] and $\sigma_{v_0} = 10$ [m/s]. Under these conditions, what is, approximately, the expected maximum height the ball will reach?

1.39 How Many Trials Produced a Given Maximum?

Let $\mathbf{N}$ be a positive discrete rv with probability distribution $p(n)$ and associated probability generating function $g(z) = \mathsf{E}[z^{\mathbf{N}}]$ (cf. page 2). Also, let $\mathbf{X}_i$ be continuous i.i.d. rvs, with a common DF F. The rv defined by

$$\mathbf{M} = \max_{i=1,\ldots,\mathbf{N}} \{\mathbf{X}_i\}$$

is then a *random maximum* — the largest of a random number (namely, **N**) of rvs (namely, the $\mathbf{X}_i$). It is intuitively clear that **M** and **N** are positively related — if **N** is large, then **M** will, on average, tend to be larger, because it is then the maximum of a larger number of i.i.d. realizations.

a. Consider the simple case that **N** equals (with probability $\frac{1}{2}$) either 1 or 2. This implies that $\mathsf{E}[\mathbf{N}] = 1.5$. Assume the $\mathbf{X}_i$ to be uniform rvs on $[0, 1]$. This means that the random maximum **M** is (with equal probability) either simply $\mathbf{X}_1$ (namely, if **N** = 1) or the maximum of $\mathbf{X}_1$ and $\mathbf{X}_2$ (i.e., if **N** = 2). Given that **M** = 0.9, what is the conditional expectation of **N**?

b. Show that in general

$$r(x) \;=\; \mathsf{E}[\,\mathbf{N}\mid \max(\mathbf{X}_1, \ldots, \mathbf{X_N}) = x\,] = 1 \;+\; F(x) \cdot \frac{g''[F(x)]}{g'[F(x)]}$$

c. Consider $r(x)$ for the case that **N** has a geometric distribution, so that $\mathsf{P}(\mathbf{N} = n) = p \cdot (1 - p)^{n-1}$.

d. Consider $r(x)$ the special case in which **N** equals either 1 or $k > 1$, both with probability $\frac{1}{2}$. For a uniform rv, $F(x) = x$, $0 \leq x \leq 1$, the case $k = 2$ corresponds to part a.

1.40 Waiting for Success

The probability p of a certain event is usually estimated by looking at how often it occurs in n independent trials. If this frequency is k, then the usual estimate of p is k/n. In this procedure, n is fixed in advance, independent of the outcome of the individual trials.

An alternative way to estimate p is to look at how long (i.e., how many trials) it takes to achieve a preset number r of successes. With this procedure, the total number of trials required is an rv, $\mathbf{N}_r$. Intuitively, the larger $\mathbf{N}_r$ is, the smaller will be our estimate of p.

a. Use $\mathsf{E}[\mathbf{N}_1]$ to derive a moment estimate, say $\hat{p}$, for p. In Problem 1.22 we already saw that $\hat{p}$ is also the ML estimate of p derived from $\mathbf{N}_1$. Find the expectation of this estimate. Is $\hat{p}$ biased?

b. Argue that $\mathsf{E}[\mathbf{N}_r] = r \cdot \mathsf{E}[\mathbf{N}_1]$, and derive a moment estimate for p based on $\mathbf{N}_r$. Use the results from Problem 1.37 to derive its approximate expectation. When will this approximation be valid?

c. Haldane[2] proposed the improved estimate $\hat{p} = (r-1)/(\mathbf{N}_r-1)$ for $r > 1$. What is the bias of Haldane's estimate?

---

[2] HALDANE, J.B.S. (1945).

2

Hints

2.1 To Begin or Not to Begin?

Consider the cases $k = 0, 1, 2, 3$ by drawing a tree diagram that depicts the possible sequence of events. Then insert the conditional probabilities along the branches of this diagram. For example, for $k = 2$ the first ball may be black ($p = 2/3$) or red ($p = 1/3$), etc.

2.2 A Tournament Problem

a. Consider one of the two female players: there are nine other players who could be selected as her opponent, and one of these nine is the other female player.

b. Consider how many different orders (that is, different with respect to the sex of the players) of the $2n$ players can be formed when k of them are females (F) and $2n - k$ are males (M). For example, for $k = 3, n = 4$, one specific such order is $< MFMMFMFM >$. It is a "positive" case because no two females will have to play against each other. How many such positive cases are there altogether?

2.3 Mean Waiting Time for $1 - 1$ vs. $1 - 2$

a. Assume that Peter starts with a 1 but fails to get another 1 in his next throw. This implies that the second throw cannot be the start of a $1 - 1$ run. Consider why this is different for Paula.

b. Conditioned on the first throw, define suitable states describing the current state of the game, such as the starting state or "last throw a 1."

2.4 How to Divide up Gains in Interrupted Games

a. Consider that Peter only needs 2 more points, whereas Paula needs 4. If they were to continue from that state on, how often would Peter have won 2 more rounds before Paula had won 4 more rounds?

b. Consider that if Peter needs to win a more rounds and Paula needs to win b more rounds for an overall win, they could pretend that they play a *fixed* number of $a + b - 1$ further rounds. Explain why with this fixed number of additional rounds *exactly one* of the two will necessarily reach the required number of wins.

2.5 How Often Do Head and Tail Occur Equally Often?

a. Compute the relevant probability using the binomial distribution.

b. Consider that for increasing n the binomial distribution converges to the normal distribution.

2.6 Sample Size vs. Signal Strength

Derive, separately for Peter and Paula, the probability of the two scenarios, conditional on $\mathcal{A}$ being red vs. blue. For example, how likely is it to draw six times a red ball, when $\mathcal{A}$ is red vs. blue? Then use Bayes' theorem.

2.7 Birthday Holidays

Let $f(n)$ be the expected number of man-days when n workers are employed. Consider f for extreme cases such as $n = 1, n = 2$, and $n \to \infty$.

2.8 Random Areas

a. Consider the case that $\mathbf{U}$ is equal to 1 or 2 with probability $\frac{1}{2}$. Then Peter's area will necessarily be either $1^2 = 1$ or $2^2 = 4$ whereas that of Paula could be 1, 2, or 4. Determine the probabilities associated with these different outcomes, and then compute the expected areas.

b. Remember that the variance is defined as $\mathsf{Var}[\mathbf{U}] = \mathsf{E}[\mathbf{U}^2] - \mathsf{E}^2[\mathbf{U}] \geq 0$, that is, $\mathsf{E}[\mathbf{U}^2] \geq \mathsf{E}^2[\mathbf{U}]$.

2.9 Maximize Your Gain

a. In order to win a large amount, one would like to choose a large value of c. On the other hand, the larger c chosen, the smaller the probability to win anything at all gets. Denote as $G(c)$ the expected gain as a function of c. Derive G by conditioning on whether $\mathbf{U} \le c$ or $\mathbf{U} > c$. Next, maximize G as a function of c.

b. The hazard function of $\mathbf{U}$ is defined as $f(u)/[1-F(u)]$; cf. page 2. Relate this to part a.

2.10 Maximize Your Gain When Losses Are Possible

a. Derive G by conditioning on whether $\mathbf{U} \le c$ or $\mathbf{U} > c$. Next, maximize G as a function of c.

b. Consider whether $G > 0$ for at least some values of c.

c. The solution of the equation

$$2e^{-x}(1 - x) = 1$$

is $x = 0.315$.

2.11 The Optimal Level of Supply

a. Consider the two extreme options $c = 0$ and $c = \infty$. What is the behavior of G as $c \to 0$ from the right?

b. Argue that conditional on $\mathbf{D} = d$ the expected gain will be equal to $d(s - b) - (c - d)b$ for $0 \le d \le c$, and equal to $c(s - b)$ for $d > c$.

2.12 Mixing RVs vs. Mixing Their Distributions

a. Consider how likely it is that $\mathbf{A}$ assumes the value of 100. How likely is this for $\mathbf{B}$? Also, consider what would happen to $\mathbf{A}$ and $\mathbf{B}$ if the variance of $\mathbf{X}$ and of $\mathbf{Y}$ would tend to zero.

b. and c. For $\mathbf{A}$, apply the general rules to obtain the expectation and variance of a sum of rvs.

For $\mathbf{B}$, introduce an indicator variable $\mathbf{I}$ that equals 1 with probability p and 0 with probability $1 - p$. Then write

$$\mathbf{B} = \mathbf{I} \cdot \mathbf{X} + (1 - \mathbf{I}) \cdot \mathbf{Y}$$

Now apply the general rules to find the expectation and variance of **B**. For example, if two rvs are independent then the expectation of their product is equal to the product of their expectations. Also, if two rvs are independent then functions of them are independent as well.

2.13 Throwing the Same vs. Different Dice

a. Denote a success (= the throwing of a 5) in any given trial as 1 and a failure as 0. Consider that the successive trials are always independent. This means that the variance of the total number of successes is the sum of the variance of the success (= 0 vs. 1) within each individual trial. Observe also that whereas the trials are always *independent*, their success probability is not *identical*: that for the trials 1-72 differs from that for trials 73-144.

b. Reformulate the question in terms of Problem 1.12: the number of successes is a "mixture" (with mixing weights $p = 1 - p = \frac{1}{2}$) of choosing either die A or die B.

c. See the hints for part a.

2.14 Random Ranks

a. Consider first the case that Peter draws just $n = 1$ value. How likely is it that Paula's draw will occupy rank 1 or rank 2? Next consider $n = 2$, enumerate all possible orders of Peter's two draws and the single draw of Paula. How many of them are such that Paula's draw is smallest, intermediate, or largest?

b. Try to generalize the reasoning in a.

2.15 Ups and Downs

a. Note that this problem really asks for $P(\mathbf{C} > \mathbf{B} \mid \mathbf{B} > \mathbf{A})$. Rewrite this according to the definition of a conditional probability, and express it in terms of the six (equally likely) possible rank orders of $(\mathbf{A}, \mathbf{B}, \mathbf{C})$.

b. Do the assumptions needed in part a. still hold for exchangeable rvs?

c. Distribute the covariances in $\mathrm{Cov}(\mathbf{B} - \mathbf{A}, \mathbf{C} - \mathbf{B})$. Determine similarly, e.g., $\mathrm{Var}(\mathbf{B} - \mathbf{A})$.

2.16 Is 2X the Same as $X_1 + X_2$?

a. Use general linearity properties of the variance; specifically, consider (and compare) the variance of $X_1 + X_2$ and of $2X$.

b. Consider first the case of $n = 2$. Then try to relate the case of $n = 2^{k+1}$ to the case $n = 2^k$.

2.17 How Many Donors Needed?

Condition on the amount x (which has an exponential density with mean 20) given by the first donor, and distinguish the cases that it is smaller or larger than $t = 100$. If $x \geq t$, then only one donor was required. If $x < t$, then the "residual problem" remains to collect a further amount of $t - x$.

2.18 Large Gaps

The Poisson process is defined by independent exponential interevent times. Look up the two properties of exponential rvs stated in Problem 1.27. Condition on the point of time when the first event occurs, and distinguish the cases that this first event occurs before or after τ.

2.19 Small Gaps

The Poisson process is defined by independent exponential interevent times. Look up the two properties of exponential rvs stated in Problem 1.27. Consider the course of events that could happen during the τ time units following the first event.

2.20 Random Powers of Random Variables

a. Consider first the special cases of $(\alpha = 0, \beta > 0)$ and $(\alpha > 0, \beta = 0)$. Condition on $M = m$, then relate to the probability generating function of N, which is defined as $g(z) = E[z^N]$.

b. Consider again the case of $\alpha = 0$.

2.21 How Many Bugs Are Left?

Assume that Peter and Paula independently detect any given error with probabilities p_A, p_B, respectively, and denote the (unknown) total number of errors in the text as n. Denote as n_A the number of errors detected by Peter, as n_B the number of errors detected by Paula, and as n_{AB} the number of errors detected by both. Derive the expected frequencies corresponding to n_A, n_B, and n_{AB} and relate them to p_A, p_B, and n.

2.22 ML Estimation with the Geometric Distribution

a. Write down the likelihood of the data, and maximize its logarithm. Consider the special case of $n = 1$.

b. The asymptotic variance of $\hat{p}$ depends on the true value of p; it is given by the negative reciprocal of the expected Fisher information,

$$\text{a.s.e.}^2(\hat{p}) = \frac{-1}{\mathsf{E}\left[\frac{\partial^2 \ln L}{\partial p^2}\right]}$$

c. Consider the Taylor expansion of $f(x) = 1/x$ at the point $x_0 = 1/p$. Also, note that as n increases a.s.e.$(\hat{p})$ tends to zero so that $\hat{p}$ will then be in a neighborhood close to p.

2.23 How Many Twins Are Homozygotic?

a. Derive the probability for a twin pair to be mm, ff, or mf. Then choose α and β so as to maximize the likelihood of the observed data.

b. If L_0, L_1 are the likelihoods of the restricted ($\beta = \frac{1}{2}$) and unrestricted (β free) model, respectively, then

$$\lambda = -2 \ln \frac{L_0}{L_1}$$

is χ^2-distributed, with df equal to the number of additional parameters determined by the restricted model, i.e., $df = 1$ in the present case.

2.24 The Lady Tasting Tea

a. Set up a 2×2 table with the true state (T vs. M) of the cups as one variable and the response the lady gives ("T" vs. "M") as the second variable. Reformulate the conditions of the experiment in terms of this table's margins. Assume the lady designates k type T cups correctly as "T", and consider what this implies for the other entries of the table.

b. Condition first on the event that the lady correctly identifies x_1 type T and x_2 type M cups, $0 \leq x_1, x_2 \leq k$. This leaves her with a total of $2n - x_1 - x_2$ nonidentified cups: $n - x_1$ of them being type T, $n - x_2$ being type M. Next, she needs to randomly designate $k - x_1$ of the $n - x_1$ type T cups as being of type T in order to obtain a total of k hits among the n cups she selects as type T. Finally, sum across the pairs (x_1, x_2) and weigh by the probabilities associated with them.

c. Insert into the result from part b., using $\theta = 0.50$.

2.25 How to Aggregate Significance Levels

a. Express the observed p-value (if H_0 is true) obtained from a single application of the test in terms of F.

b. If U is a uniform rv on $[0, 1]$, then the rv $V = -2 \ln U$ has an exponential distribution with rate $\lambda = \frac{1}{2}$, i.e., a χ^2-distribution with $df = 2$. This implies that the sum of two independent V_i will have a χ^2-distribution with $df = 4$.

c. In the aggregation procedure suggested by b., which pairs of (p_1, p_2) will lead to a significant value of χ_4^2?

2.26 Approximately How Tall Is the Tallest?

a. Note that the largest of n independent realizations of U is smaller than m if and only if all n realizations are individually smaller than m. That is, $P(M_n \leq m) = [P(U \leq m)]^n$. Also note that the rv $F(U)$ is uniformly distributed. The DF of the uniform distribution is $H(z) = z$, and therefore

$$P[F(U) \leq 1 - x/n] = P[U \leq F^{-1}(1 - x/n)] = 1 - x/n$$

b. Use the fact that if n is not too small $(1 - x/n)^n \approx e^{-x}$; then apply the results from part a.

c. For the exponential distribution, $1 - F(x) = \exp(-\lambda x)$. Insert this into the results from part b.

d. Use part b. with the cumulative normal distribution, $\Phi(x)$. An excellent approximation[1] to its extreme tail $(x > 5.50)$ is given by

$$1 - \Phi(x) \approx \frac{1}{x\sqrt{2\pi}} \cdot \exp\left(-\frac{x^2}{2} - \frac{0.94}{x^2}\right)$$

2.27 The Range in Samples of Exponential RVs

a. Consider the development of the process "chronologically." Also, first look at the range in the simplest case of $n = 2$.

b. Apply the lack-of-memory property of exponential rvs as described in the statement of the problem: conditional on $\mathbf{T}_{(1)} = t$, the remaining $n - 1$ rvs all have a displaced exponential distribution, with their origin at t.

2.28 The Median in Samples of Exponential RVs

a. For $\mathbf{M}$ to equal m, how many realizations have to be smaller than, equal to, and larger than m?

b. Using the two properties of exponential rvs described in Problem 1.27, consider the expected distances from one order statistic to the next.

2.29 Breaking the Record

a. Consider the case that the $n = 100$ successive realizations are arranged in perfectly increasing (or perfectly decreasing) order.

Let $\mathsf{E}(n)$ be the expected number of records with n realizations. Try to explicitly derive $\mathsf{E}(n)$ for $n = 1, 2, 3, 4$. Then relate $\mathsf{E}(n)$ to $\mathsf{E}(n-1)$, distinguishing the two cases that $\mathbf{X}_n$ is, or is not, larger than the largest value seen so far. How likely are these two cases?

Which effect would a monotone increasing transformation have that is applied to all n realizations?

b. Condition on the value of $\mathbf{X}_1$.

c. Show that if the proposition holds for r then it also holds for $r + 1$.

d. Look up the lack-of-memory property stated in Problem 1.27. What does this imply about the excess by which each new record surpasses the previous record? Insert the exponential distribution into the general result for g_r found in part c.

---

[1] Due to DERENZO, S.E. (1977).

2.30 Paradoxical Contribution

a. Consider the separate roles of the two means, of the two variances, of the correlation, and of the fact that 175 g is considerably above the sum of the individual averages. Note that the question really asks for a particular conditional expectation. The expectation of exactly which conditional distribution is this?

b. Should the expectation for bird A now be smaller, equal to, or larger than with the scenario in a.?

2.31 Attracting Mediocrity

a. Given the measurement of 105 and the value of $\sigma = 3$, how likely is it that Peter or Paula was selected? For example, if Peter was selected, which result would you predict for the second measurement?

More formally, use Bayes' theorem to find the probability that Peter or Paula had been selected, given the value of 105 observed. Then weigh your predictions for the second measurement by these conditional probabilities.

b. Consider extreme special cases such as $\sigma \to 0$ or $\sigma \to \infty$.

c. Note that a given measurement can be represented as $\mathbf{X} = \mathbf{T} + \mathbf{E}$, the sum of the (randomly selected) true IQ, plus the associated (positive or negative) measurement error. If we knew the true IQ $\mathbf{T}$ of the specific person selected, then we should clearly predict the value $\mathbf{T}$ as the expected outcome of the second measurement of that person (given that the error of the second measurement has a mean of zero). Then, what we are looking for is $\mathsf{E}[\mathbf{T}|\mathbf{T} + \mathbf{E} = x]$. This is a special case of our solution to Problem 1.30.

2.32 Discrete Variables with Continuous Error

a. Consider extreme cases for σ, such as $\sigma \to 0$ and $\sigma \to \infty$.

b. Condition on $\mathbf{N} = n$, then write down the explicit formula for f.

2.33 The High-Resolution and the Black-White View

a. Write down the likelihood function $L(\mu)$, take its natural logarithm, and maximize it.

Under wide conditions (satisfied here)

$$\text{a.s.e.}^2(\hat{\mu}) = \frac{-1}{\mathsf{E}\left[\frac{\partial^2 \ln L}{\partial \mu^2}\right]}$$

b. Formulate the likelihood function $L(\mu)$ for the data as seen by researcher B. Take its natural logarithm, and maximize. One intuition here might be that k/n represents an estimate of the probability for a single measurement not to exceed the threshold c; this probability equals $1 - \exp(-c/\mu)$.

c. Use the general expression for the a.s.e. stated earlier. Also, think about the qualitative dependence of the a.s.e. on c. For example, what happens for very small and very large values of c?

d. Minimize the result obtained in c. with respect to the threshold c; compare to the result obtained in a.

2.34 The Bivariate Lognormal

a. Start from the definition of the variance; see page 1. Then use the fact that $[\exp(a)]^2 = \exp(2a)$. Relate this to the given expectation of $\mathbf{U}$.

b. Recall that functions of independent rvs are independent as well.

c. Note that scaling parameters in the (linear) relation between $\mathbf{X}$ and $\mathbf{Y}$ turn into powers in the relation between $\mathbf{U}$ and $\mathbf{V}$.

d. Start from the definition of the covariance; to deal with the term $\mathsf{E}[\mathbf{UV}]$, note that $\exp(a) \cdot \exp(b) = \exp(a + b)$. Then relate again to the result about the expectation of $\mathbf{U}$.

2.35 The arcsin($\sqrt{p}$) Transform

a. Note that if n is large, then $\hat{p}$ is necessarily close to p. Thus, Δp is bound to be small, so that Taylor's approximation holds in a neighborhood of the order of Δp around p (cf. Problem 1.37).

b. Consider the variance of $g(\hat{p})$ derived in part a. and specify which condition the function g must satisfy so that this variance becomes independent of p. Specifically, consider in this context the differential equation

$$g'(p) = \frac{1}{2\sqrt{p(1-p)}}$$

2.36 Binomial Trials Depending on a Latent Variable

If n is not too small, then $\hat{p}$ will usually be relatively close to the true population value, p. This means that $\hat{c} = \Phi^{-1}(\hat{p})$ will also be close to the true population value, c. Then use Taylor's approximation to Φ^{-1} around p.

2.37 The Delta Technique with One Variable

a. Consider the Taylor expansion of $f(x)$ around μ up to the order 2. Then take the expectation with respect to $\mathbf{X}$.

b. Note that for $f(x) = e^x$, we have $f(x) = f'(x) = f''(x)$. Therefore, the second-order Taylor expansion of e^x around $x_0 = \mu$ is

$$f(x) \approx e^\mu \left[1 + (x - \mu) + \frac{1}{2}(x - \mu)^2 \right]$$

Now determine the approximation to $\mathsf{E}[f(\mathbf{X})]$.

c. From elementary mechanics, for a given angle α the maximum height is equal to

$$f(\alpha) = \frac{v_0^2}{2g} \sin^2(\alpha)$$

where g is the gravitational constant and v_0 is the initial velocity.

2.38 The Delta Technique with Two Variables

a. Form the Taylor expansion of $f(x, y)$ around the point (μ_x, μ_y) up to the order 2. Then take the expectation with respect to $(\mathbf{X}, \mathbf{Y})$.

b. Insert the special case of $f(x, y) = x/y$ into the result of part a., using the facts that the second derivatives of f with respect to x and y are zero and $2x/y^3$, respectively.

c. First note that $f(x, y)$ may be written in product form as $g(x) \cdot h(y)$, where $g(x) = x$ and $h(y) = 1/y$. Functions (such as g, h) of independent rvs are themselves independent; thus, the expectation of their product factors into the product of their individual expectations.

d. This problem addresses a special case of the situation considered in parts b. and c. Also, recall that a χ^2–rv with r degrees of freedom has expectation r and variance $2r$.

e. Look up the solution to part c. of Problem 1.37. Then apply the results from part a. to the function

$$f(v_0, \alpha) = \frac{v_0^2}{2g} \sin^2(\alpha)$$

2.39 How Many Trials Produced a Given Maximum?

a. Consider the conditional probability $P[N = 1 \,|\, M = 0.9]$. Apply Bayes' formula to it, and note that $P[N = 2 \,|\, M = 0.9]$ is necessarily the complement thereof. Then form $E[N \,|\, M = 0.9]$.

b. To find the conditional expectation $E[N \,|\, \max(X_1, \ldots, X_N) = x]$, first take a closer look at the associated conditional distribution, the expectation of which we seek to derive. Thus, use Bayes' theorem to relate the known quantity $P[\max(X_1, \ldots, X_N) = x \,|\, N = n]$ to the quantity we are looking for, namely, $P[N = n \,|\, \max(X_1, \ldots, X_N) = x]$. Next use this latter distribution to get an expression for its expected value. Finally simplify this expression, using elementary manipulations of the generating function g.

c. The generating function in this case is

$$g(z) = \frac{pz}{1 - z(1 - p)}$$

d. The generating function in this case is

$$g(z) = \frac{z + z^k}{2}$$

2.40 Waiting for Success

a. The waiting time N_1 has a geometric distribution. Also, make use of the identity $(-1 \le x < 1)$

$$\sum_{n=1}^{\infty} \frac{1}{n} x^n = -\ln(1 - x)$$

b. The rv N_r is distributed as is the sum of r independent rvs N_1. The latter rv has mean $1/p$ and variance $(1 - p)/p^2$. Therefore, N_r has mean r/p and variance $r \cdot (1 - p)/p^2$. Consider the results from Problem 1.37 for the case of $f(x) = r/x$.

c. Note that the rv N_r has the so-called negative-binomial distribution,

$$p_r(n) = \binom{n - 1}{r - 1} p^r (1 - p)^{n-r} \quad , \text{ for } n \ge r.$$

3

Solutions

3.1 To Begin or Not to Begin?

The best strategy depends on the parity of k. If k is even then the player who starts drawing has an advantage, but if k is odd, Peter's and Paula's chances are equal.

To see why, consider a slightly revised, but functionally equivalent procedure to play this game: first Peter and Paula draw balls as before, but without inspecting their color, continuing until the urn is empty. Each player then inspects the balls in front of him or her, and the player who finds the red ball in his or her own sample has won. Notice that with this revised procedure the winner will always be the same as with the original procedure. Therefore, the probability of winning for each player will also be the same as in the original game.

Consider first the case that k is odd, so that the total number of balls, $1 + k$, is even. Thus, after all balls are removed, both players will have drawn an equal number of balls, namely $\frac{1+k}{2}$. Now, each half of the $1+k$ balls has the same probability to contain the red one. Thus, the probability for Peter and Paula to find the red ball contained in his or her own sample is $\frac{1}{2}$. However, when k is even, then the total number $1 + k$ of balls in the urn is odd. Thus, the player who starts will eventually have sampled $1 + \frac{k}{2}$ balls, as opposed to only $\frac{k}{2}$ balls drawn by the other player. Thus, the ratio favoring the starting player is

$$\frac{1 + \frac{k}{2}}{\left(1 + \frac{k}{2}\right) + \frac{k}{2}} = \frac{2 + k}{2 + 2k}$$

For example, if $k = 0$, the player who starts will necessarily always win. If $k = 2$, the starting player collects a total of two balls out of three, and so has a probability of $\frac{2}{3}$ to win. Therefore, Paula should generally start drawing: if k is even, she gains a real advantage, and if k is odd at least nothing is lost.

3.2 A Tournament Problem

Let us say a given player is at rank i if his or her name is the ith one drawn from the urn. The total number of different ways to distribute the k female players among the $2n$ ranks is $\binom{2n}{k}$. Clearly, not all of them are favorable; for example, if a female player is drawn first (rank 1) and another female player is drawn second (rank 2), then the condition mentioned in the problem is not satisfied.

How many different *favorable* ways are there to distribute the k females players among the $2n$ ranks? Imagine we put together successive pairs of ranks, namely $(1,2),(3,4),\ldots,(2n-1,2n)$. Of these n pairs we may select k to associate exactly one female player with each of them; the number of ways to select these k pairs from a total of n pairs is $\binom{n}{k}$. However, within each of the k pairs chosen, the respective female may be placed either at the odd or at the even rank of that pair, yielding a total of $2^k \cdot \binom{n}{k}$ different ways to distribute k female players among $2n$ ranks such that each of the n pairs has at most one female.

$p(k,n)$ then is the ratio of the favorable over all possible cases, i.e.,

$$p(k,n) = \frac{2^k \cdot \binom{n}{k}}{\binom{2n}{k}} = \frac{\binom{2n-k}{n}}{\binom{2n}{n}} \cdot 2^k$$

Note that the formula also covers the trivial cases of $k = 0, 1$.

For a constant fraction of female players $p(k,n)$ declines rapidly as n increases. For example, with $k = 2$ females among $2n = 10$ players, chances are $8/9 = 88.9\%$ that there will be no match involving only females (i.e., that those two females will not have to play against each other). However, with $k = 20$ females among $2n = 100$ players (thus, again a fraction of 20% females), the corresponding probability drops to about 9.2%. That is, with a probability of more than 90% there will among all 50 matches be at least one match with two female players.

3.3 Mean Waiting Time for $1 - 1$ vs. $1 - 2$

a. Peter's sequence will usually consist of a certain number of 1s, each of which is followed by a number different from 1 with $p = 5/6$. If, after obtaining a 1, he fails to achieve the desired $1 - 1$ run, then the number thrown was necessarily different from 1 and, therefore, cannot constitute the potential beginning of a potential $1 - 1$ run. This is different for Paula: if she fails to throw a 2 after an initial 1, then she *may* do so by throwing another 1, which in turn could be the start of a potential $1 - 2$ run. Therefore, the expected waiting time will be somewhat shorter for Paula.

b. As for Peter, let us assume that in each throw a 1 occurs with a fixed probability of p. We are looking for the expected waiting time to see two 1s in a row for the first time; call this expectation W.

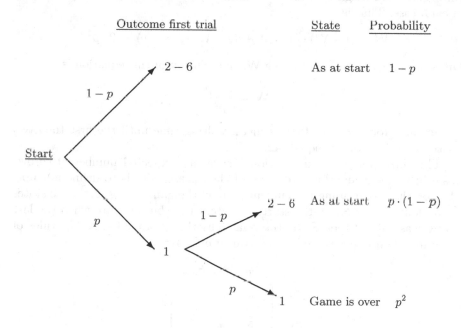

Fig. 3.1. Tree diagram showing the possible starting events of the game and their probabilities. The insert "$2 - 6$" stands for any of the integers from 2 to 6.

Consider the outcome of Peter's first throw. With probability $1 - p$, the outcome is different from 1, that is, one of the numbers from 2 to 6. In this case (shown as the upper branch in Figure 3.1) one trial is used up, and the state of the game after the first trial is as it was before the game started. This means that the *additional* expected waiting time — namely, in addition to the first trial — is just as large as the original expected waiting time was; that is, it equals W. Therefore, if the first case applies (which has probability $1 - p$), then the conditional expectation of the total waiting time is equal to $1 + W$.

With probability p, Peter's first trial results in a 1, in which case we need to consider the outcome of the second trial. If his second trial results in a 1 too, then he has achieved two 1s in a row, and the game is over after only two trials. This case is shown as the lowest branch in Figure 3.1, and its overall probability is equal to p^2. Alternatively, his second trial might not yield another success, in which case he is again in the same state as he was right before the start of the game, except that he has already spent two trials. This case is shown as the middle branch in Figure 3.1. The expected number of throws in this case is $2 + W$, and its overall probability is equal to $p \cdot (1 - p)$.

These three cases exhaust all opening possibilities, and they are mutually exclusive. Therefore, we can obtain an equation for W by conditioning on these three cases and weighing each conditional expectation of W by the probability of each case. That is,

$$W = (1 + W) \cdot (1 - p) + (2 + W) \cdot p(1 - p) + 2 \cdot p^2$$

On isolating the terms containing W, the solution of this equation is

$$W = \frac{1 + p}{p^2}$$

Specifically, for $p = 1/6$, Peter's mean waiting time until the first time two consecutive 1s show up equals 42.

The corresponding computation for Paula's expected number of throws is slightly less straightforward. One of the easiest ways is to define a simple Markov chain describing the current state of the game using just three states S_i, as follows. Let S_0 be the starting state, let S_1 be the state when the last throw was a 1, and let S_2 be the final (absorbing) state $1 - 2$. The rules of the game then imply the following transition matrix:

$$
\begin{array}{c}
 \\
S_0 \\
 \\
S_1 \\
 \\
S_2
\end{array}
\begin{array}{ccc}
S_0 & S_1 & S_2 \\
\left[\begin{array}{ccc}
\frac{5}{6} & \frac{1}{6} & - \\
 & & \\
\frac{2}{3} & \frac{1}{6} & \frac{1}{6} \\
 & & \\
- & - & 1
\end{array}\right]
\end{array}
$$

For example, a throw in state S_0 (at start) will lead with probability 1/6 into state S_1 (namely, if a 1 is thrown), and with probability 5/6 will Paula stay in state S_0 (with any other number thrown). Similarly, after a 1 has been thrown (i.e., out of state S_1), Paula will either stay in state S_1 (if a further 1 is thrown, probability 1/6), or she will enter the final state S_2 (if a 2 is thrown, probability 1/6), or she will enter state S_0 (any number other than 1 or 2, probability 2/3).

Therefore, if W_i denotes the mean number of steps to the absorption at S_2 out of state S_i, then reasoning much as we did to derive W for Peter

$$W_0 = 1 + \frac{5}{6}W_0 + \frac{1}{6}W_1$$
$$W_1 = 1 + \frac{2}{3}W_0 + \frac{1}{6}W_1 + \frac{1}{6}W_2$$

(note that by definition $W_2 = 0$). For example, out of state S_0, the expected number of throws needed to get into S_2 consists in any case of the throw done in this state, plus a further number of throws needed thereafter. This further number equals W_1 if the first throw results in a 1 (probability 1/6), and it equals W_0 if the first throw results in some number other than 1 (probability 5/6).

Solving for W_0, we get $W_0 = 36$. Thus, Paula will on average need only 36 throws, whereas Peter requires on average 42. What at first appears to be a trivial and negligible variation of the rules has in fact a noticeable influence on how many throws will be needed.

3.4 How to Divide up Gains in Interrupted Games

a. Peter and Paula seem not to have agreed on how to proceed when the game is interrupted before one of them has won 5 points. Clearly, then, one option is that both simply retain their original $50 because the game has not been completed according to the rules originally agreed on.

On the other hand, Peter might argue that he had already made considerable progress toward winning 5 points and therefore claim more than his original $50. A rational basis of this claim could be to consider in how many similar cases Peter would finally have won over Paula — if the game had continued. More precisely, Peter needs 2 more points, whereas Paula still needs twice as many, namely 4. Thus we may legitimately ask: what is the probability that Peter would have won the 2 required points before Paula had won 4 points?

The hard way to compute this probability is to enumerate all possible sequences that end favorably for Peter and to sum their probabilities. Note that Peter's victory necessarily ends with a point made by him and may be preceded by 1, 2, 3, or at most 4 rounds of which Peter has won exactly one. The complete list contains 10 possible sequences of successive winners (A = Peter wins a round, B = Paula wins a round): { AA, ABA, BAA, ABBA, BABA, BBAA, ABBBA, BABBA, BBABA, BBBAA }, and their probabilities are easily found to add up to $26/32 = 0.8125$. This result suggests that Peter might claim $81.25, in which case Paula receives $18.75. Clearly, when the number of rounds that Peter and Paula would still need to win gets larger, then this procedure of explicit enumeration becomes fairly laborious.

A more elegant way to go about this problem is to realize that after at most 5 more rounds either Peter or Paula must have won: if Peter wins 2 or more of these 5 rounds, then Paula has won at most 3 rounds — not enough for her to achieve the required total of 5 points. Conversely, if Paula wins 4 or 5 rounds, Peter has won at most 1. Therefore, by an argument somewhat similar to the one we used to solve Problem 1.1, a procedure equivalent to the original rules is to pretend to play *exactly* 5 more rounds: if Peter wins 2 or more of these, this is equivalent to him reaching a total of 5 points before Paula did. On the other hand, if Paula wins 4 or more of these 5 rounds, then she must have reached a total of 5 points before Peter has. The advantage of this reconceptualization is that for a fixed number of rounds (namely, 5) the required probabilities are easily computed from the binomial distribution with $n = 5$, $p = \frac{1}{2}$. For example, the probability that Peter wins no round or only 1 round out of 5 is

$$\binom{5}{0}\left(\frac{1}{2}\right)^5 + \binom{5}{1}\left(\frac{1}{2}\right)^5 = \frac{6}{32}$$

in agreement with the result obtained by the lengthy explicit enumeration described earlier.

b. More generally, if Peter needs to win a more rounds and Paula needs to win b more rounds for an overall win, they could pretend that they play a *fixed* number of $a + b - 1$ further rounds. Exactly one of the following two mutually exclusive events will then obtain:

1. Peter wins a or more points, in which case Paula wins less than b, or
2. Paula wins b or more points, in which case Peter wins less than a.

The first event implies that Peter had won a points before Paula had won b; the second event means that Paula had won b points before Peter had won a. Therefore, with respect to the winning probabilities, the procedure is functionally equivalent to the rules originally set out, namely to stop when either Peter has collected a, or when Paula has collected b further points.

The probability, e.g., of the first event enumerated here is then easily found by summing the respective binomial terms:

$$P(\text{Peter wins overall} \,|\, a, b) = \frac{1}{2^{a+b-1}} \cdot \sum_{i=a}^{a+b-1} \binom{a+b-1}{i}$$

3.5 How Often Do Head and Tail Occur Equally Often?

a. From the binomial distribution, the probability of an even $n : n$ split with $2n$ coins equals

$$p(2n) = \binom{2n}{n} \cdot 2^{-2n}$$

For $2n = 20$ coins, the probability $p(20) = 0.1762$, meaning that in expectation only in 17.62% of all experiments in which 20 coins are thrown, will there be an even 10 : 10 split. This value is already considerably smaller than the even-split probability with just two coins, which equals $p(2) = \frac{1}{2}$.

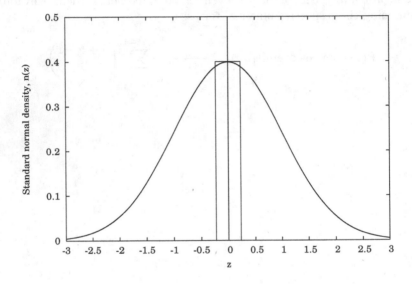

Fig. 3.2. The area under the standard normal density between $z_l = -1/\sqrt{2n}$ and $z_u = 1/\sqrt{2n}$ can be approximated by a rectangle of width $\sqrt{2/n}$ and height $1/\sqrt{2\pi}$, the value of the normal density at $x = 0$. The figure illustrates this approximation for $n = 10$.

b. For increasing n, the direct computation of $p(2n)$ gets impractical and does not easily permit more general insights. However, we know that the binomial converges to the normal distribution as n increases. Given that the mean and variance of the binomial distribution with $2n$ trials and $p = \frac{1}{2}$ are $\mu = n$ and $\sigma^2 = n/2$, this fact implies that $p(2n)$ must tend to the area under the standard normal density from the lower limit

$$z_l = \frac{(n - 0.5) - n}{\sqrt{n/2}} = -\frac{1}{\sqrt{2n}}$$

to the upper limit

$$z_u = \frac{(n+0.5) - n}{\sqrt{n/2}} = \frac{1}{\sqrt{2n}}$$

For example, $z_u = -z_l = 0.2236$ for the value $n = 10$ considered earlier, which yields the approximation 0.1769, very close to the exact value.

As is indicated in Figure 3.2, this area under the standard normal density may in turn be further approximated geometrically by a rectangle of width $dz = z_u - z_l = \sqrt{2/n}$ and height equal to the ordinate of the standard normal density at the midpoint between z_l and z_u, which is $z_m = 0$. From the equation of the standard normal density, we know that the ordinate at $z_m = 0$ is equal to $1/\sqrt{2\pi}$. Referring to Figure 3.2,

$$p(2n) \approx (\text{width of rectangle}) \times (\text{height of rectangle})$$
$$= \sqrt{2/n} \cdot \frac{1}{\sqrt{2\pi}}$$
$$= \frac{1}{\sqrt{n\pi}}$$

a function that decreases monotonically with the square root of n.

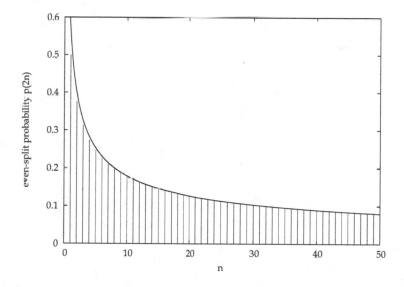

Fig. 3.3. Exact binomial even-split probabilities $p(2n)$ and the approximation $p(2n) = 1/\sqrt{n\pi}$ for an equal number (n) of heads and tails when $2n$ coins are thrown.

For example, for $n = 10$, this approximation yields 17.84%, as compared to the exact result of 17.62% derived in a. The geometry of Figure 3.2 indicates that the area of the approximating rectangle overestimates the actual area under the normal density, but even for $n = 10$ this error is quite small. For $n = 100$, we get $p(2n) = 3.99\%$, and for $n = 1000$, we have $p(2n) = 1.78\%$. Figure 3.3 shows as vertical lines the exact binomial probabilities and as a smooth curve the approximation $p(2n) = \frac{1}{\sqrt{n\pi}}$, which is excellent. Clearly, then, as n increases, the probability of an even split gets smaller and smaller, contrary to intuitive notions.

As so often, our intuition is perhaps better characterized as fuzzy rather than wrong. Indeed, our considerations imply that the *relative* deviations of the two frequencies — namely, relative to the number of coins thrown — will decrease steadily as n increases. This means, for example, that the two *relative* frequencies will both approach 50% closer and closer as n increases. But, as Figure 3.3 suggests, this does not mean that the (absolute) frequencies themselves will differ by less and less.

3.6 Sample Size vs. Signal Strength

Many people believe that Peter's evidence for $\mathcal{A}$ being blue is much stronger than that of Paula. In fact, their evidence is exactly equally strong. What matters in the present situation is only the *difference* of the number of red and blue balls drawn; this difference being given, it is unimportant on how many observations (draws) it was based. Here are the relevant computational details.

Let r, b stand for red and blue, respectively. What we really want to know in this problem is this: what is the probability of $\mathcal{A}$ being blue if (e.g.) six out of six balls drawn are red? Correspondingly, what is the probability of $\mathcal{A}$ being blue, if 303 out of 600 balls drawn are red?

Consider the first question. The conditional probability that we are looking for here is $\mathsf{P}(\mathcal{A} = b| < rrrrrr >)$, where $< rrrrrr >$ stands for a sequence of six successive draws of red balls. By Bayes' theorem,

$$\mathsf{P}(\mathcal{A} = b| < rrrrrr >) =$$

$$= \frac{\mathsf{P}(< rrrrrr > |\mathcal{A} = b)\mathsf{P}(\mathcal{A} = b)}{\mathsf{P}(< rrrrrr > |\mathcal{A} = b)\mathsf{P}(\mathcal{A} = b) + \mathsf{P}(< rrrrrr > |\mathcal{A} = r)\mathsf{P}(\mathcal{A} = r)}$$

$$= \frac{1}{1 + \frac{\mathsf{P}(<rrrrrr>|\mathcal{A}=r)}{\mathsf{P}(<rrrrrr>|\mathcal{A}=b)}}$$

because $\mathsf{P}(\mathcal{A} = b) = \mathsf{P}(\mathcal{A} = b) = \frac{1}{2}$, so that this factor cancels throughout. Also, if $\mathcal{A} = b$ then three red and two blue balls remained in the urn, so that in each independent draw a red ball had a probability of 3/5 to be selected. Hence, $\mathsf{P}(< rrrrrr > |\mathcal{A} = b) = (3/5)^6$. Similarly, $\mathsf{P}(< rrrrrr > |\mathcal{A} = r) = (2/5)^6$. Putting these facts together and canceling common terms, we see that

$$\mathsf{P}(\mathcal{A} = b| < rrrrrr >) = \frac{1}{1 + (2/3)^6} = 0.919$$

Thus, conditional on the six observations made by Peter, the ball $\mathcal{A}$ was blue with about 91.9%.

Now let us look at Paula's data. By the binomial distribution, if $\mathcal{A}$ was blue then her probability to obtain, in 600 draws, 303 red and 297 blue balls is equal to

$$\mathsf{P}(303\, r \wedge 297\, b|\mathcal{A} = b) = \binom{600}{303} \left(\frac{3}{5}\right)^{303} \left(\frac{2}{5}\right)^{297}$$

In a similar way,

$$\mathsf{P}(303\, r \wedge 297\, b|\mathcal{A} = r) = \binom{600}{303} \left(\frac{2}{5}\right)^{303} \left(\frac{3}{5}\right)^{297}$$

The ratio of the latter to the former probability is $(2/3)^6$; thus, for Paula we obtain

$$P(\mathcal{A} = b | 303\, r \wedge 297\, b) = \frac{1}{1 + (2/3)^6}$$

This value is identical to that for Peter, meaning that their evidence in favor of $\mathcal{A}$ being blue is actually equally strong.

Obviously, the same will be true whenever the difference between red and blue balls drawn is the same for Peter and Paula. Given that difference — the "signal strength" — the number of draws in which it was obtained does not differentially influence the strength of the evidence obtained.

As stated, many people tend to think that Peter's evidence for $\mathcal{A}$ being blue is much stronger. In fact, the large conditional probability of nearly 92% is quite in line with this intuition. What seems much more surprising though is that the conditional probability is equally large for Paula. To see why, it is important to realize that a split of 303/297 is in fact quite unlikely under *both* hypotheses, $\mathcal{A} = b$ and $\mathcal{A} = r$. For example, in the first case we would expect about 360 red and about 240 blue balls among 600 draws; in the second case these numbers would be reversed. Thus, the split of 303/297 is — in absolute terms — actually quite unlikely under both hypotheses, but — in relative terms — it is much more compatible with the hypothesis of $\mathcal{A} = b$ than with $\mathcal{A} = r$.

3.7 Birthday Holidays

Let $f(n)$ be the expected number of man-days when n workers are employed. We seek to maximize f as a function of n. Thus, first we need to derive an explicit expression for f. Even before we do so, it is easy to see that there must be at least one maximum for the function f. Clearly, $f(1) = 364$, and neglecting the probability of a birthday coincidence, $f(2) \approx 2 \cdot 363 = 726$. Thus, initially f increases. On the other hand, as $n \to \infty$, there will be no birthday-free day at all, meaning that the expected number of man-days tends to zero as n gets very large. Together, then, f must have at least one maximum.

To locate this maximum, consider first some specific day i, where $i = 1, \ldots, 365$. Let $\mathbf{X}_i$ be an rv that equals 1 if none of the n workers has his birthday on this specific day whereas $\mathbf{X}_i = 0$ if at least one worker has his birthday on that day. With n workers, how likely is it that $\mathbf{X}_i = 1$? Clearly, for any individual worker, the probability not to have his birthday on day i equals $364/365$. Therefore, with independent birthdays, the probability $P(\mathbf{X}_i = 1)$ that no worker has his birthday on day i is equal to $(364/365)^n$. This implies that the expectation $\mathsf{E}[\mathbf{X}_i] = P(\mathbf{X}_i = 1) = (364/365)^n$.

Now, define the rv $\mathbf{S} = \sum_{i=1}^{365} \mathbf{X}_i$. In words, $\mathbf{S}$ is the total number of birthday-free days. Clearly, the $\mathbf{X}_i$ are not independent, but for the expectation $\mathsf{E}[\mathbf{S}]$ we have

$$\mathsf{E}[\mathbf{S}] = \mathsf{E}[\sum_{i=1}^{365} \mathbf{X}_i]$$

$$= \sum_{i=1}^{365} \mathsf{F}[\mathbf{X}_i]$$

$$= 365 \cdot (364/365)^n$$

Now, $f(n) = n \cdot \mathsf{E}[\mathbf{S}]$, because in expectation the n workers will have to work on $\mathsf{E}[\mathbf{S}]$ days, yielding $n \cdot \mathsf{E}[\mathbf{S}]$ man-days. Thus,

$$f(n) = n \cdot 365 \cdot (364/365)^n$$

Figure 3.4 shows the function f. Looking at the change of the sign of the difference $f(n+1) - f(n)$ it is easy to see that its maxima occur at $n = 364$ and $n = 365$, yielding $f(364) = f(365) = 48943.5$ expected man-days. In the case of $n = 364$, each worker needs to work, in expectation, on 134.5 days per year; in the case of $n = 365$ this expectation drops to 134.1 days per year. On the level of the man-days, this drop of man-days per worker is just compensated by the additional man-days contributed by the added 365th worker.

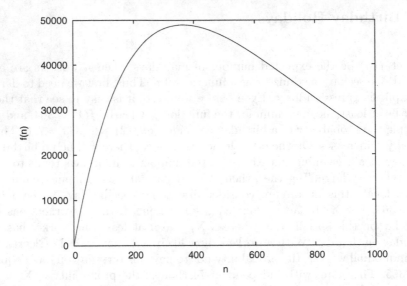

Fig. 3.4. The expected number, $f(n)$, of man-days per year if n workers are employed. The maxima occur at $n = 364$ and $n = 365$, yielding $f(364) = f(365) = 48943.5$ expected man-days. With these values, each worker will have to work about 134 days per year.

3.8 Random Areas

Peter just generates a single realization of $\mathbf{U}$, and so the area of his square will be equal to $\mathbf{U}^2$. Thus, on average the area of his square is equal to $A_{Peter} = \mathsf{E}[\mathbf{U}^2]$.

Paula generates two independent realizations of $\mathbf{U}$ for the width and length of her rectangle, let us call these realizations $\mathbf{U}_1$ and $\mathbf{U}_2$. Accordingly, the area of her rectangle will be equal to $\mathbf{U}_1 \cdot \mathbf{U}_2$. On average the area of her rectangle equals

$$
\begin{aligned}
A_{Paula} &= \mathsf{E}[\mathbf{U}_1 \cdot \mathbf{U}_2] \\
&= \mathsf{E}[\mathbf{U}_1] \cdot \mathsf{E}[\mathbf{U}_2] \\
&= \mathsf{E}^2[\mathbf{U}]
\end{aligned}
$$

because $\mathbf{U}_1$ and $\mathbf{U}_2$ are independent and have the same distribution as $\mathbf{U}$.
    Therefore,

$$
A_{Peter} - A_{Paula} = \mathsf{E}[\mathbf{U}^2] - \mathsf{E}^2[\mathbf{U}] = \mathsf{Var}[\mathbf{U}] \geq 0
$$

that is, $A_{Peter} \geq A_{Paula}$. On average, the area of Peter's square is larger than that of Paula's rectangle, even though all lengths and widths of all rectangles (a square is a rectangle) are generated by realizations of the same generic rv, $\mathbf{U}$.

3.9 Maximize Your Gain

a. As stated in the Hints, in order to win a large amount, one would like to choose a rather large value of c. On the other hand, the larger c is chosen, the smaller the probability of winning anything at all.

Let us call the chosen value of c the "strategy" of the player and denote its expected gain as $G(c)$. Thus, with the strategy c, the gain will be zero with probability $P(\mathbf{U} \leq c) = F(c)$, and it will be equal to c with probability $P(\mathbf{U} > c) = 1 - F(c)$. Putting these two cases together, we get

$$G(c) = 0 \cdot F(c) + c \cdot [1 - F(c)] = c \cdot [1 - F(c)]$$

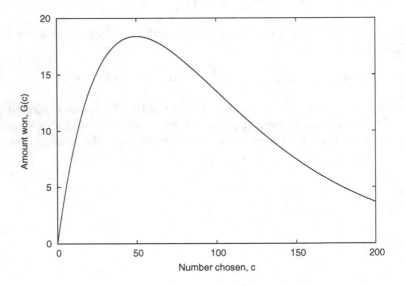

Fig. 3.5. The relation between the expected gain, $G(c)$, and the number, c, that was chosen. The example assumes that $\mathbf{U}$ has an exponential distribution $f(t) = \lambda \exp(-\lambda t)$ with mean $1/\lambda = 50$. The optimal choice of c is then just 50, in which case the expected gain equals 18.39.

To get an idea about G, it is useful to first study its boundary values. From the "tail formula" (see page 2) for the expected square of $\mathbf{U}$,

$$E[\mathbf{U}^2] = \mu^2 + \sigma^2$$

$$= 2 \int_0^\infty u[1 - F(u)] \, du$$

$$= 2 \int_0^\infty G(u) \, du \; < \; \infty$$

because the first two moments of **U** were assumed to be finite. It must then be the case that $G(c)$ tends to zero as $c \to \infty$; otherwise the integral would diverge. Also, G is nonnegative, and from $G(0) = G(\infty) = 0$ we conclude that G must have at least one maximum over the positive reals.

Differentiating G with respect to c and setting this derivative equal to zero gives

$$\frac{d\,G(c)}{dc} = [1 - F(c)] - cf(c) = 0$$

i.e., $$\frac{f(c)}{1 - F(c)} = \frac{1}{c}$$

b. The left-hand side of the last equation is the hazard function h of **U**; cf. page 2. The equation states that the optimal strategy c is the abscissa at which the hazard function $h(c)$ crosses the function $1/c$.

c. For an exponential rv, the hazard function is constant, $h(c) = \lambda$. Therefore, the optimal strategy is $c = 1/\lambda$, i.e., to choose the expected value of **U**, see Figure 3.5. With this strategy, the expected gain is

$$G(\frac{1}{\lambda}) = \frac{1}{\lambda} \cdot e^{-1}$$

i.e., about 36.8% of the expected value of **U**. This value appears remarkably small, but due to the large spread (relative to its mean) of the exponential distribution, there is no way to further improve it within the given setting.

3.10 Maximize Your Gain When Losses Are Possible

a. With the strategy c, the gain (actually, loss) will be equal to $-c$ with probability $P(U \leq c) = F(c)$, and it will be equal to $+c$ with probability $P(U > c) = 1 - F(c)$. Weighing these two cases by their respective probabilities, we get

$$G(c) = -c \cdot F(c) + c \cdot [1 - F(c)]$$
$$= c \cdot [1 - 2F(c)]$$

Figure 3.6 shows the qualitative behavior of G.

Differentiating G with respect to c and setting this derivative equal to zero yields the optimum condition

$$\frac{d\,G(c)}{dc} = 1 - 2\,[cf(c) + F(c)] = 0$$

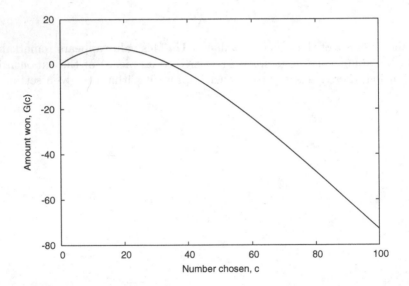

Fig. 3.6. The relation between the expected gain $G(c)$ and the number c that was chosen. The example assumes that U has an exponential distribution $f(t) = \lambda \exp(-\lambda t)$, with mean $1/\lambda = 50$. The optimal choice of c is then 15.75, in which case the expected gain equals 7.24.

b. Note from the form of $G(c)$ that if c is smaller than the median of U, then $G(c)$ is positive. Therefore, the global maximum of G cannot be negative.

c. For an exponential rv with rate λ, the optimum condition yields

$$2e^{-\lambda c}\left(1 - \lambda c\right) = 1$$

The solution to this equation is $c = 0.315/\lambda$. Compared to Problem 1.9, the less favorable rules of the current game clearly suggest a rather more cautious strategy; compare Figure 3.6 to Figure 3.5. Correspondingly, the expected gain with the optimal strategy decreases to just $0.145/\lambda$, roughly 40% of what the player could expect under the rules of the previous problem. On the other hand, although less attractive, the game should still be accepted by the player, because its expected gain is positive when the optimal strategy is chosen.

3.11 The Optimal Level of Supply

a. Clearly, if he orders no milk at all, there will be no gain and no loss, $G(0) = 0$. For very small c, the man will almost certainly sell his entire lot c, thereby gaining $c(s-b)$. Thus, for small c, the net gain G will increase linearly with slope $s - b$ and thus be positive. For very large c, almost surely not all stocked milk will be sold; in this situation, each additional unit stocked will impose an additional loss of b. Thus, for c large, G will decrease linearly with slope $-b$. Therefore, G must have at least one positive maximum for $c > 0$.

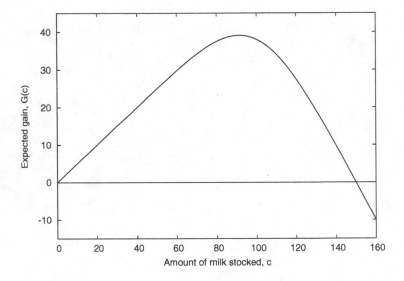

Fig. 3.7. The typical relation between expected net gain, $G(c)$, and the amount of milk stocked, c. The example assumes that one unit of milk costs $b = 1$ and is sold for $s = 1.50$. The demand is normally distributed, with mean 100 and standard deviation 20. (The normal distribution is not strictly positive, but for the values chosen it practically is.) The optimal amount the milkman should stock is then about 91 units, in which case his expected net gain is 39.1.

b. If he orders c units and sells only some of them, $0 \le d \le c$, then the conditional expected net gain $G(c|\mathbf{D} = d)$ will be equal to $d(s - b) - (c - d)b$ because his gain from the d units sold is $d(s-b)$, and the cost of the remaining $c - d$ nonsold units is $(c - d)b$. On the other hand, if $d \ge c$ he will sell all c units, yielding $G(c|\mathbf{D} = d) = c(s - b)$.

Conditioning on the demand $\mathbf{D}$ we get

$$G(c) = \int_0^\infty G(c|\mathbf{D} = d) \cdot f(d)\, dd$$

$$= \int_0^c [d(s-b) - (c-d)b]f(d)\, dd + \int_c^\infty c(s-b)f(d)\, dd$$

$$= s \cdot \int_0^c [1 - F(d)]\, dd - cb$$

after slight rearrangements and an integration by parts. From the usual necessary maximum condition

$$\frac{dG(c)}{dc} = s[1 - F(c)] - b = 0$$

We then get

$$c_{opt} = F^{-1}\left(1 - \frac{b}{s}\right)$$

where F^{-1} is the inverse function of F; it is unique because F was assumed to be strictly increasing. Note that

$$\frac{d^2 G(c)}{dc^2} = -sf(c) < 0$$

so that G is concave downward: starting from $(0|0)$ it first increases toward a unique maximum at c_{opt} and then decreases, with an asymptotic slope of $-b$. The typical qualitative behavior of the function G is shown in Figure 3.7.

3.12 Mixing RVs vs. Mixing Their Distributions

a. The densities of **A** and **B** are shown in Figure 3.8.

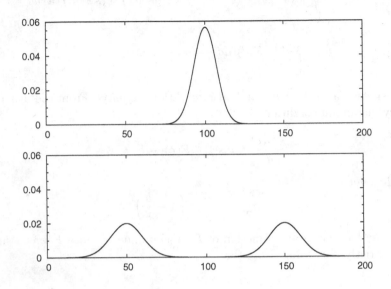

Fig. 3.8. Upper panel: density of the mixture of two normal rvs, lower panel: mixture of two normal densities.

Note that **A** will in each realization be close to the mean of 100. In contrast, **B** will in each realization be close to either 50 or 150 but will practically never be equal to its mean of 100.

b and c. Let us first consider the general case, and then in turn insert the specific results for $p = \frac{1}{2}$.

We start with the rv

$$\mathbf{A} = p \cdot \mathbf{X} + (1 - p) \cdot \mathbf{Y}$$

where **X** and **Y** are independent rvs. On taking expectations and variances we immediately get from the usual standard rules for the moments of sums of independent rvs

$$E[\mathbf{A}] = p \cdot E[\mathbf{X}] + (1 - p) \cdot E[\mathbf{Y}]$$

and

$$\mathrm{Var}[\mathbf{A}] = p^2 \cdot \mathrm{Var}[\mathbf{X}] + (1 - p)^2 \cdot \mathrm{Var}[\mathbf{Y}]$$

For example, if the variances of both **X** and **Y** tend to zero, then the variance of **A** will tend to zero as well.

In the normal distribution example, first note that $p \cdot \mathbf{X}$ and $(1 - p) \cdot \mathbf{Y}$ are both normal rvs, too. Therefore, $\mathbf{A}$ is itself normally distributed because it is the sum of two normal rvs. As shown in the top panel of Figure 3.8, its mean is 100 and its variance is 50. The standard deviation of slightly above 7 means that the interval $100 \pm 2 \cdot 7 = [86, 114]$ brackets approximately 95% of the distribution of $\mathbf{A}$.

To obtain the corresponding results for $\mathbf{B}$, we introduce a binary variable $\mathbf{I}$ that indicates from which of the two distributions the current realization of $\mathbf{B}$ was drawn. Specifically, $\mathbf{I} = 1$ with probability p and $\mathbf{I} = 0$ with probability $1 - p$. Clearly, $\mathbf{I}$ and $\mathbf{I}^2$ take on the same values (namely, 0 and 1) with the same probabilities: thus, $\mathsf{E}[\mathbf{I}] = \mathsf{E}[\mathbf{I}^2] = p$. This implies that $\mathsf{Var}[\mathbf{I}] = \mathsf{E}[\mathbf{I}^2] - \mathsf{E}^2[\mathbf{I}] = p(1 - p)$.

Using $\mathbf{I}$, we can write

$$\mathbf{B} = \mathbf{I} \cdot \mathbf{X} + (1 - \mathbf{I}) \cdot \mathbf{Y}$$

Now we apply the standard rules to determine the expectation and variance of $\mathbf{B}$. For the expectation we get

$$\begin{aligned} \mathsf{E}[\mathbf{B}] &= \mathsf{E}[\mathbf{I} \cdot \mathbf{X}] + \mathsf{E}[(1 - \mathbf{I}) \cdot \mathbf{Y}] \\ &= \mathsf{E}[\mathbf{I}] \cdot \mathsf{E}[\mathbf{X}] + \mathsf{E}[(1 - \mathbf{I})] \cdot \mathsf{E}[\mathbf{Y}] \\ &= p \cdot \mu_x + (1 - p) \cdot \mu_y \end{aligned}$$

By the definition of the variance

$$\mathsf{Var}[\mathbf{B}] = \mathsf{E}[\mathbf{B}^2] - \mathsf{E}^2[\mathbf{B}]$$

We have already determined the expectation of $\mathbf{B}$, so we next turn to the expectation of its square. On squaring $\mathbf{B}$

$$\mathbf{B}^2 = \mathbf{I}^2 \cdot \mathbf{X}^2 + (1 - \mathbf{I})^2 \cdot \mathbf{Y}^2 + 2 \cdot \mathbf{I} \cdot \mathbf{X} \cdot (1 - \mathbf{I}) \cdot \mathbf{Y}$$

we have, factoring the expectation of independent rvs (such as $\mathbf{I}^2$ and $\mathbf{X}^2$):

$$\mathsf{E}[\mathbf{B}^2] = \mathsf{E}[\mathbf{I}^2] \cdot \mathsf{E}[\mathbf{X}^2] + \mathsf{E}[(1 - \mathbf{I})^2] \cdot \mathsf{E}[\mathbf{Y}^2] + 2 \cdot \mathsf{E}[\mathbf{I}(1 - \mathbf{I})] \cdot \mathsf{E}[\mathbf{X}] \cdot \mathsf{E}[\mathbf{Y}]$$

The required insight at this stage is that $\mathbf{I}(1 - \mathbf{I})$ necessarily equals zero (as $\mathbf{I}$ equals zero or one). Thus, $\mathsf{E}[\mathbf{I}(1 - \mathbf{I})] = 0$, so that the last summand drops out altogether. This leaves us with

$$\mathsf{E}[\mathbf{B}^2] = p \cdot \mathsf{E}[\mathbf{X}^2] + (1 - p) \cdot \mathsf{E}[\mathbf{Y}^2]$$

On the other hand, on squaring $\mathsf{E}[\mathbf{B}]$ we get

$$\mathsf{E}^2[\mathbf{B}] = p^2 \cdot \mathsf{E}^2[\mathbf{X}] + (1 - p)^2 \cdot \mathsf{E}^2[\mathbf{Y}] + 2p(1 - p) \cdot \mathsf{E}[\mathbf{X}] \cdot \mathsf{E}[\mathbf{Y}]$$

Forming $\mathsf{E}[\mathbf{B}^2] - \mathsf{E}^2[\mathbf{B}]$ and simplifying we obtain our final expression

$$\text{Var}[\mathbf{B}] = p \cdot \sigma_x^2 + (1-p) \cdot \sigma_y^2 + p(1-p) \cdot (\mu_x - \mu_y)^2$$

Comparing with the result for $\text{Var}[\mathbf{A}]$ we immediately see that $\text{Var}[\mathbf{A}] < \text{Var}[\mathbf{B}]$ if we exclude the trivial cases of $p = 0$ and $p = 1$. It is instructive and characteristic that whereas the variance of $\mathbf{A}$ equals zero if the variances of $\mathbf{X}$ and of $\mathbf{Y}$ are zero, the same is not true for the rv $\mathbf{B}$ (except for the trivial case that the expectations of $\mathbf{X}$ and $\mathbf{Y}$ are also the same).

For the case of the two normals, the mean of $\mathbf{B}$ equals 100, just as the mean of $\mathbf{A}$. However, its variance and standard deviation are very much larger than that of $\mathbf{A}$; specifically, $\sigma_B = 51$.

3.13 Throwing the Same vs. Different Dice

a. In this problem the successive trials are independent, but the first and the last 72 trials have a different success probability.

Let $X_i, i = 1, \ldots, 144$, equal 1 if a success occurs in trial i and be equal to zero if not. With the sampling scheme in a., the probability of obtaining a success during the trials $1, \ldots, 72$ is equal to $\frac{1}{4}$, thus $E[X_i] = P(X_i = 1) = \frac{1}{4}$ and $Var[X_i] = P(X_i = 0) \cdot P(X_i = 1) = \frac{3}{16}$. The success probability during the trials $73, \ldots, 144$ is equal to $\frac{1}{12}$, thus $E[X_i] = P(X_i = 1) = \frac{1}{12}$ and $Var[X_i] = P(X_i = 0) \cdot P(X_i = 1) = \frac{11}{144}$.

Let $S = \sum_{i=1}^{144} X_i$ be the total number of successes during all 144 trials. The X_i are not all identically distributed; they are, however, independent of each other, so that the variance of their sum is equal to the sum of their individual variances. That is,

$$Var[S] = 72 \cdot \left(\frac{3}{16} + \frac{11}{144} \right)$$
$$= 19$$

Thus, although the mean number of successes is equal to that in the standard binomial scheme (namely, 24), the variance around that mean is actually somewhat *smaller*.

This same effect would be even more pronounced if in the first 72 trials the success probability were $p = \frac{1}{3}$, and in the last 72 trials $p = 0$, yielding again a mean success probability of $p = \frac{1}{6}$ across all 144 trials. On average, there would thus again be 24 successes, but the variance would only be equal to $72 \cdot \frac{2}{9} - 16$. Many people find this counterintuitive, because the variation of the basic success probability across trials seems to "throw in additional variance," but that is not true.

b. It is important to realize that in this problem the successive trials are not independent. For example, if the first trial yielded a success, it gets more likely that the second trial will also yield a success, because a success in the first trial represents probabilistic evidence that die A was chosen. This argument would become very obvious if die A had a success probability of $p = \frac{1}{3}$, and die B of $p = 0$ (the average of both dice being again $\frac{1}{6}$). Then a success in the first trial necessarily implies that die A must have been chosen.

The present problem is actually a variant of the one treated in Problem 1.12. Let $\mathbf{I}$ be a binary rv that takes (with probability $\frac{1}{2}$) the value 1, which indicates that die A was selected, and (with probability $\frac{1}{2}$) the value 0, indicating that die B was selected. Using the rv $\mathbf{I}$ we may write

$$\mathbf{S} = \mathbf{I} \cdot \mathbf{X} + (1 - \mathbf{I}) \cdot \mathbf{Y}$$

where $\mathbf{X}$ and $\mathbf{Y}$ are the numbers of success with die A and B, respectively. Obviously, $\mathbf{X}$ is a binomial rv with $n = 144$ and $p = \frac{1}{4}$, and $\mathbf{Y}$ is a binomial rv with $n = 144$ and $p = \frac{1}{12}$. This implies that the mean and variance of $\mathbf{X}$ are $\mu_x = 144 \cdot \frac{1}{4} = 36$ and $\sigma_x^2 = 144 \cdot \frac{1}{4} \cdot \frac{3}{4} = 27$. Similarly, the mean and variance of $\mathbf{Y}$ are $\mu_y = 144 \cdot \frac{1}{12} = 12$ and $\sigma_y^2 = 144 \cdot \frac{1}{12} \cdot \frac{11}{12} = 11$. In Problem 1.12 we have seen that

$$\mathsf{E}[\mathbf{S}] = \frac{1}{2} \cdot \mu_x + \frac{1}{2} \cdot \mu_y = 24$$

and

$$\mathsf{Var}[\mathbf{S}] = \frac{1}{2} \cdot \sigma_x^2 + \frac{1}{2} \cdot \sigma_y^2 + \frac{1}{4} \cdot (\mu_x - \mu_y)^2 = 163$$

Thus, the variance in the present scheme is much larger than in the standard binomial case with $p = \frac{1}{6}$. The reason is that the choice for die A or B is made only once, for all 144 throws. If we select die A, we expect on average a total of 36 successes, whereas if we select die B, we will on average just see 12 successes. The mean number of successes remains 24, but the spread around this mean gets very large under the present sampling scheme. Depending on whether we choose die A or B, the number of successes will typically range from about 5 to about 46 — a very considerable variation!

c. In this problem the successive trials are again independent because the "current" die is selected anew each time, independent of all previous events, and as all trials are executed in the same way, they all must have the same success probability. Clearly this success probability equals $(\frac{1}{4} + \frac{1}{12})/2 = \frac{1}{6}$ in each individual trial, and so the situation corresponds exactly to the usual binomial scheme. Therefore, the mean and variance of the number of successes are 24 and 20, respectively.

Why is the variance with the present scheme slightly higher than in scheme a.? Intuitively, it is important to realize that in scheme a. it is *guaranteed* that exactly 72 trials are done with die A and with die B. In contrast, with the present sampling scheme, the actual number of trials carried out with die A and with B will fluctuate somewhat around 72, thereby adding a further variance component.

3.14 Random Ranks

a. Consider first the scenario when Peter draws just two values; let us call these values a and b. Let us call x the value obtained by Paula. Clearly, because the draws are independent and are obtained from the same population, all six orders are equally likely: $abx, axb, bax, bxa, xab, xba$, where, e.g., bxa stands for the order $b < x < a$. Two of them are such that Paula's draw (x) is first, second, or third. Thus, all three ranks for Paula's draw are equally likely.

b. In general, there are $(n+1)!$ equally likely orders (i.e., permutations) of the n values drawn by Peter and the single value drawn by Paula. How many of them are such that Paula's value ends up at rank k? If rank k is reserved for Paula's value then there remain $n!$ orders in which the n values drawn by Peter can be arranged around this rank. For example, if $n = 3$, and Paula's value (call it x) is fixed at rank 2, Peter's three values (say, a, b, c) can be arranged as follows: $axbc, axcb, bxac, bxca, cxab, cxba$. These are those six outcomes in which Paula's value x ends up at rank no. 2. Obviously, exactly the same generative procedure could be used to put Paula's value x at any rank from 1 to 4. In general, then, of the $(n+1)!$ permutations there are $n!$ such that Paula's value appears at rank no. k. Thus, the probability for each possible rank k of Paula's draw is equal to $\frac{n!}{(n+1)!} = \frac{1}{(n+1)}$. Note that this value is independent of the actual rank, k.

Specifically, with $n = 100$ each rank for Paula's draw from 1 to 101 is equally likely. Intuitively, most people think that it is much more likely that Paula's draw will eventually occupy some middle, intermediate rank, such as $k = 51$, rather than an extreme rank such as $k = 1$ or $k = 101$. But that is not correct: any rank is equally likely.

3.15 Ups and Downs

a. Consider the six possible rank orders of $(\mathbf{A}, \mathbf{B}, \mathbf{C})$; each of these orders is equally likely because the rvs were assumed to be identically and independently distributed. If, e.g., abc stands for the increasing order $\mathbf{A} < \mathbf{B} < \mathbf{C}$, then three of the six orders are such that $\mathbf{A} < \mathbf{B}$, so that an increment occurs with $\mathbf{B}$. These are those orders in which the letter a occurs before b, namely, abc, acb, cab. The mere fact that there are three (an odd number) such orders already implies that following an increment with $\mathbf{B}$ there cannot be an equal probability for increments and decrements with $\mathbf{C}$. Indeed, only the case abc yields a further increment, whereas acb and cab represent a decrement, i.e., $\mathbf{C} < \mathbf{B}$. Therefore, the conditional probability of a decrement with $\mathbf{C}$, given an increment at $\mathbf{B}$ is $\frac{2}{3}$: despite the independence of all rvs involved, ups are more often followed by downs than by a further up.

b. Let $f(a, b, c)$ denote the joint density of $(\mathbf{A}, \mathbf{B}, \mathbf{C})$. The assumptions stated in question b. imply that, e.g., $f(a, b, c) = f(c, a, b)$, and similarly for any permutation of the three arguments. Therefore, each of the six possible rank orders must still be equally likely, so that the argument from a. still goes through, even for dependent but exchangeable rvs.

c. To find this correlation, we first determine the associated covariance

$$
\begin{aligned}
\mathsf{Cov}(\mathbf{B} - \mathbf{A}, \mathbf{C} - \mathbf{B}) &= \mathsf{Cov}(\mathbf{B}, \mathbf{C}) - \mathsf{Var}(\mathbf{B}) - \mathsf{Cov}(\mathbf{A}, \mathbf{C}) + \mathsf{Cov}(\mathbf{A}, \mathbf{B}) \\
&= \gamma^2 - \sigma^2
\end{aligned}
$$

where γ^2 denotes the (common) covariance between any pair of the three rvs. Also, from standard linearity properties of the variance we find

$$
\begin{aligned}
\mathsf{Var}(\mathbf{B} - \mathbf{A}) &= \mathsf{Var}(\mathbf{B}) + \mathsf{Var}(\mathbf{A}) - 2\,\mathsf{Cov}(\mathbf{B}, \mathbf{A}) \\
&= 2\left(\sigma^2 - \gamma^2\right)
\end{aligned}
$$

and for reasons of symmetry, $\mathsf{Var}(\mathbf{C} - \mathbf{B}) = 2\left(\sigma^2 - \gamma^2\right)$. Therefore, the correlation

$$
\begin{aligned}
\mathsf{corr}(\mathbf{B} - \mathbf{A}, \mathbf{C} - \mathbf{B}) &= \frac{\mathsf{Cov}(\mathbf{B} - \mathbf{A}, \mathbf{C} - \mathbf{B})}{\sqrt{\mathsf{Var}(\mathbf{B} - \mathbf{A}) \cdot \mathsf{Var}(\mathbf{C} - \mathbf{B})}} \\
&= \frac{\gamma^2 - \sigma^2}{2(\sigma^2 - \gamma^2)} \\
&= -\frac{1}{2}
\end{aligned}
$$

The successive differences $(\mathbf{B} - \mathbf{A})$ and $(\mathbf{C} - \mathbf{B})$ are negatively correlated because $\mathbf{B}$ occurs in both expressions, but with opposite signs. For example, if $\mathbf{B}$ is particularly large, then $(\mathbf{B} - \mathbf{A})$ will usually be positive and $(\mathbf{C} - \mathbf{B})$ will usually be negative.

3.16 Is 2X the Same as $X_1 + X_2$?

a. Let σ^2 denote the variance of **X**. According to the general rules, the variance of the rv $2\mathbf{X}$ is $4\sigma^2$, whereas the variance of the rv $\mathbf{X}_1 + \mathbf{X}_2$ is equal to $2\sigma^2$. Therefore, if both of these rvs are to follow the same — Cauchy — distribution, then σ^2 must be either infinite or zero. However, in the latter case the probability mass would be concentrated at a single point, and so **X** would be a constant, not a Cauchy rv. Therefore, the first case applies, and $\sigma^2 = \infty$.

b. Consider first the case of $n = 2$. If, as stated, $\mathbf{X}_1 + \mathbf{X}_2$ is distributed as is $2\mathbf{X}$, then it is also true that **X** is distributed as is $(\mathbf{X}_1 + \mathbf{X}_2)/2$, i.e., the arithmetic mean of a sample of two. Thus, the property holds for $n = 2^k = 2$, i.e. for $k = 1$.

Next, assume that the sum of 2^k independent Cauchy rvs is distributed as is $2^k\mathbf{X}$. Consider then the sum of 2^{k+1} independent Cauchy rvs: it may be thought of as 2^k independent Cauchy rvs plus another 2^k independent Cauchy rvs. By assumption, both components are individually distributed as is $2^k\mathbf{X}$, and so their sum has the same distribution as $2^k(\mathbf{X}_1 + \mathbf{X}_2)$. But we saw earlier that $\mathbf{X}_1 + \mathbf{X}_2$ is distributed as is $2\mathbf{X}$, and so the sum of 2^{k+1} independent Cauchy rvs has the same distribution as $2^k(2\mathbf{X}) = 2^{k+1}\mathbf{X}$.

By induction, then, the sum of 2^k independent Cauchy rvs is distributed as is $2^k\mathbf{X}$. Dividing by the scalar $n = 2^k$ yields the assertion: the arithmetic mean of $n = 2^k$ independent Cauchy rvs is distributed as is any of the contributing summands individually. This property seems rather counterintuitive and appears to many as deeply disturbing: it defies our nearly universal belief that the precision of arithmetic means necessarily increases with increasing sample size. What seems especially disturbing is that the Cauchy distribution (see Figure 1.1 looks fairly "well-behaved": it does not have particularly bizarre or exoctic shape and is not easily dismissed as a mere mathematical curiosity. The property actually holds for any n, not just for $n = 2^k$, but this fact requires slightly more elaborate arguments.

3.17 How Many Donors Needed?

Denote as t the amount that needs to be collected (in the present problem, $t = 100$), and let $\mathbf{A}_i$ be the amount of money given by donor $i = 1, 2, \ldots$, which has an exponential distribution with mean $1/\lambda = 20$. Also, let the "running total" be $\mathbf{S}_n = \sum_{i=1}^n \mathbf{A}_i$. The basic problem then is to find information about the discrete rv

$$\mathbf{N} = \min_{n \geq 1} \{\mathbf{S}_n \geq t\}$$

that is, the smallest index n such that $\mathbf{S}_n$ is t or more. Let $p(n|t)$, $n = 1, 2, \ldots$, denote the probability that $\mathbf{N} = n$ when a total of t needs to be raised.

Our basic approach is to condition on the amount given by the first donor, $\mathbf{A}_1$. If $\mathbf{A}_1$ is larger than t, then $\mathbf{N} = 1$; the probability $p(1|t)$ of this event is $P(\mathbf{A}_1 \geq t) = \exp(-\lambda t)$. To determine $p(n|t)$ for $n > 1$, assume that the first donor gives an amount $\mathbf{A}_1 = x$, $x < t$, which has the density $\lambda \exp(-\lambda x)$. For $\mathbf{N}$ to be equal to n, following this first donor, the residual amount of $t - x$ needs to be raised from the next $n - 1$ donors, an event that by definition has probability $p(n-1|t-x)$. The essential insight here is that when $\mathbf{A}_1 = x$, $x < t$, then we are left with a situation that is exactly analogous to the original problem, except that now we ask for the probability that $n - 1$ donors are required to first reach or exceed $t - x$ dollars.

Integrating across the possible values of $0 \leq x < t$ gives

$$p(n|t) = \int_0^t \lambda e^{-\lambda x} p(n - 1|t - x) \, dx, \quad n > 1$$

This is the basic relation that determines the $p(n|t)$, and its explicit solution is essentially a technical matter. One blunt way to solve this elementary integral equation is simply to guess the building pattern by computing the first few successive terms $p(n|t)$, using the known result for $p(1|t)$.

A more systematic way is to pass to Laplace transforms (see the definition of the moment generating function on page 2), which turn convolutions into products. Denoting the Laplace transform of $p(n|t)$ with respect to t as $L(s|n) = \int_0^\infty e^{-st} p(n|t) \, dt$,

$$L(s|n) = \frac{\lambda}{\lambda + s} \cdot L(s|n - 1), \quad n > 1$$

But from $p(1|t) = \exp(-\lambda t)$, we know that $L(s|1) = 1/(\lambda + s)$, and so by induction

$$L(s|n) = \frac{1}{\lambda} \cdot \left(\frac{\lambda}{\lambda + s}\right)^n, \quad n \geq 1$$

We note that (again using the convolution property) the last factor is just the Laplace transform of the density of the sum of n exponential rvs, each

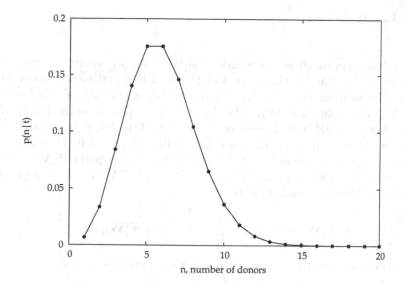

Fig. 3.9. The probability $p(n|t)$ that exactly n donors are needed to collect at least $t = 100$ dollars when the amount each donor gives is exponentially distributed, with a mean of 20 dollars. On average, it requires 6 donors, not 5, as the intuitive estimate $100/20 = 5$ would suggest.

with rate λ; therefore, its inverse transform is the gamma density (cf. page 1). Taking the factor $1/\lambda$ into account, we obtain the shifted Poisson distribution:

$$p(n|t) = \frac{(\lambda t)^{n-1}}{(n-1)!} \cdot e^{-\lambda t}$$

an expression that is also seen to be valid for $n = 1$. Thus, the number $\mathbf{N}$ of donors required to collect t or more dollars is distributed as is $1 + \mathbf{M}$, where $\mathbf{M}$ has a Poisson distribution with mean λt. From this representation, it directly follows that $\mathsf{E}[\mathbf{N}] = 1 + \lambda t$ and $\mathsf{Var}[\mathbf{N}] = \lambda t$.

For the values of $t = 100$ and $1/\lambda = 20$, the mean number of donors is 6, not 5, as the intuitive estimate $100/20 = 5$ would suggest. Figure 3.9 shows the corresponding distribution $p(n|t)$.

3.18 Large Gaps

Our basic approach is to condition on the time $\mathbf{T}_1$ when the first event occurs after the start of the process at $t = 0$. If this first event occurs after τ, then the waiting time $\mathbf{W}$ was simply τ; the probability of this event is $\exp(-\lambda\tau)$. Put more formally, $\mathsf{E}[\mathbf{W}|\mathbf{T}_1 > \tau] = \tau$. On the other hand, with density $\lambda \exp(-\lambda t)$, the first event occurs at some time $t < \tau$. In this case, the process is at t in the same state as it originally was at $t = 0$. Therefore, the *residual* waiting time, from t on, is in expectation again equal to $\mathsf{E}[\mathbf{W}]$, and so the expected total waiting time in this case ($t < \tau$) is $\mathsf{E}[\mathbf{W}|\mathbf{T}_1 = t] = t + \mathsf{E}[\mathbf{W}]$.

Putting these two cases together,

$$\mathsf{E}[\mathbf{W}] = \tau e^{-\lambda\tau} + \int_0^\tau \lambda e^{-\lambda t}\left(t + \mathsf{E}[\mathbf{W}]\right)dt$$

Simplifying and isolating the expectation, the solution is

$$\mathsf{E}[\mathbf{W}] = \frac{1}{\lambda} \cdot (e^{\lambda\tau} - 1)$$

For example, if a car passes on average every 10 seconds (i.e., $1/\lambda = 10$), and 20 seconds are needed to pass the road, then the expected waiting time is about 1 minute and 4 seconds. For a slower pedestrian who needs 40 seconds to pass the same road, the expected waiting time increases sharply to 8 minutes and 56 seconds, more than eight times as long.

3.19 Small Gaps

Clearly, to even get a chance for a coincidence, we need at least the first event to occur. In expectation, this takes $1/\lambda$ time units. Following this first event, two mutually exclusive scenarios may take place. Either the next (i.e., second) event occurs within the next τ time units — in this case, by definition, the coincidence occurs with this second event. With density $\lambda \exp(-\lambda t)$ the second event occurs $0 \le t \le \tau$ time units after the first one, and the expected total waiting time in this case is equal to $1/\lambda + t$. Or no further event occurs during the first τ time units following the first one, an event that has a probability of $\exp(-\lambda\tau)$. In this case, the state $t = \tau$ time units after the first event is exactly the same as it was originally at $t = 0$, and so the *residual* waiting time, from there on, is again $\mathsf{E}[\mathbf{U}]$, in addition to the $1/\lambda + \tau$ time units that have already elapsed. Putting these two cases together,

$$\mathsf{E}[\mathbf{U}] = \int_0^\tau \left(\frac{1}{\lambda} + t \right) \lambda e^{-\lambda t}\, dt + \left(\frac{1}{\lambda} + \tau + \mathsf{E}[\mathbf{U}] \right) e^{-\lambda\tau}$$

Simplifying and isolating the expectation, the solution is

$$\mathsf{E}[\mathbf{U}] = \frac{1}{\lambda} \cdot \frac{2 - e^{-\lambda\tau}}{1 - e^{-\lambda\tau}}$$

For example, if in a situation of exposition to a radiating substance particles are absorbed on average every 10 seconds (i.e., $1/\lambda = 10$) and a critical (e.g., damaging) event requires that two particles be absorbed within no more than $\tau = 1$ second, then the expected waiting time to the damage is about 115 seconds. If the "critical period" (the period of vulnerability) is doubled to $\tau = 2$ seconds, then the damage will in expectation occur after about 65 seconds. If, on the other hand, the mean interevent time is cut in half to $1/\lambda = 5$ seconds, then the damage will (with $\tau = 1$) in expectation occur after only less than 33 seconds, underlining the asymmetric roles played by the rate λ and by the critical period τ, even if their product is kept constant. Note that if the product $\lambda\tau$ is large — i.e., if the critical period is much larger than the mean interevent time of the Poisson process, $\tau \gg 1/\lambda$ — then the expected waiting time $\mathsf{E}[\mathbf{W}]$ tends to $2/\lambda$, reflecting the fact that under these conditions the waiting time to the coincidence simply reduces to the waiting time until the first two events have occurred.

3.20 Random Powers of Random Variables

a. As a preliminary result, let us first determine the pgf $g(z)$ of the rv $\mathbf{N}$. By the definition of a pgf (cf. page 2),

$$g(z) = E[z^{\mathbf{N}}] = \sum_{n=0}^{\infty} z^n \cdot P(\mathbf{N} = n)$$

$$= \sum_{n=0}^{\infty} z^n \cdot e^{-\beta} \beta^n / n!$$

$$= e^{-\beta} e^{z\beta} \sum_{n=0}^{\infty} e^{-z\beta} (z\beta)^n / n!$$

$$= e^{(z-1)\beta}$$

because the sum in the next-to-last line adds up the probabilities of a Poisson rv with rate $z\beta$ (cf. page 1). Note that $g(z)$ is finite for any real z, because the series $\sum_{n=0}^{\infty} x^n / n!$ converges to e^x for any real x.

Next, we condition on $\mathbf{M} = m$:

$$E[\mathbf{Q}] = \sum_{m=0}^{\infty} E[\mathbf{Q}|\mathbf{M} = m] \cdot P(\mathbf{M} = m)$$

$$= \sum_{m=0}^{\infty} E[m^{\mathbf{N}}] \cdot e^{-\alpha} \alpha^m / m!$$

Note that for any real m, the expression $E[m^{\mathbf{N}}]$ is equal to $g(m)$, as determined before. Therefore,

$$E[\mathbf{Q}] = \sum_{m=0}^{\infty} e^{(m-1)\beta} \cdot e^{-\alpha} \alpha^m / m!$$

$$= e^{-(\alpha+\beta)} \sum_{m=0}^{\infty} \left(\alpha e^{\beta} \right)^m / m!$$

$$= e^{-(\alpha+\beta)} \exp\left(\alpha e^{\beta} \right) \sum_{m=0}^{\infty} \exp\left(-\alpha e^{\beta} \right) \left(\alpha e^{\beta} \right)^m / m!$$

This sum is again of the general Poissonian form, with mean αe^{β}; it therefore adds up to 1. This gives

$$E[\mathbf{Q}] = \exp\left[\alpha \left(e^{\beta} - 1 \right) - \beta \right]$$

b. For convenience, let $E[\mathbf{Q}] = \exp[\psi(\alpha, \beta)]$; taking logarithms then gives $\psi(\alpha, \beta) = \alpha \left(e^{\beta} - 1 \right) - \beta$. Thus, $E[\mathbf{Q}]$ increases if and only if ψ increases.

Now,

$$\frac{\partial \psi}{\partial \alpha} = e^\beta - 1$$

which is positive for $\beta > 0$, since for $\beta = 0$ we have $\mathsf{E}[\mathbf{M}^0] = 1$. Therefore, $\mathsf{E}[\mathbf{Q}]$ generally increases with α.

Consider next that

$$\frac{\partial \psi}{\partial \beta} = \alpha e^\beta - 1$$

and this partial derivative is negative whenever $\alpha < e^{-\beta}$. Now, $0 < e^{-\beta} \leq 1$, so that there will be regions in which the partial derivative $\partial \psi / \partial \beta$ is negative whenever $\alpha < 1$. Figure 3.10 illustrates the principal cases using $\alpha = 0.0, 0.4$, and 1.1.

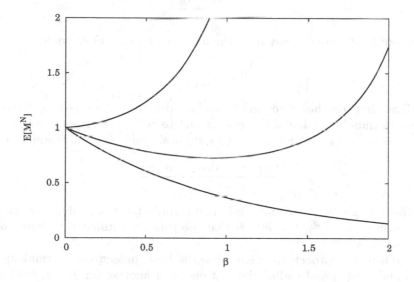

Fig. 3.10. The expectation of $\mathbf{M^N}$ as a function of β, the mean of $\mathbf{N}$, for $\alpha = 0.0$ (bottom curve), 0.4 (middle curve), and 1.1 (top curve), where α is the mean of $\mathbf{M}$.

To better understand this result, consider a special case of this condition, namely, $\alpha = 0$, in which case $\mathbf{M}$ will be equal to zero with probability 1. Now, because $0^0 = 1$, and $0^n = 0$ for $n > 0$, the expectation of $\mathbf{M^N}$ reduces to $P(\mathbf{N} = 0) = e^{-\beta}$, the only contributing term. But the larger β, the smaller will be $e^{-\beta}$, and so $\mathsf{E}[\mathbf{Q}]$ decreases monotonically with β when $\alpha = 0$ (see Figure 3.10, bottom curve). If $0 < \alpha < 1$ then $\mathsf{E}[\mathbf{Q}]$ first decreases and then increases with β (Figure 3.10, middle curve). Finally, if $\alpha \geq 1$ then $\mathsf{E}[\mathbf{Q}]$ increases monotonically with β (Figure 3.10, top curve).

3.21 How Many Bugs Are Left?

Assume that Peter and Paula independently detect any given error with probabilities p_A, p_B, and denote the (unknown) total number of errors in the text as n. Denote as n_A, n_B, and n_{AB}, respectively, the number of errors detected by Peter, by Paula, and by both. The expected number of errors found by Peter is $n \cdot p_A$, yielding the estimate $\hat{p}_A = \frac{n_A}{n}$. Similarly, $\hat{p}_B = \frac{n_B}{n}$. The expected number of errors found by both is $n \cdot p_A \cdot p_B$. Equating this expectation to the corresponding observation n_{AB} and inserting the previous estimates $\hat{p}_A, \hat{p}_B$ gives[1]

$$
\begin{aligned}
n_{AB} &= n \cdot \hat{p}_A \cdot \hat{p}_B \\
&= n \cdot \frac{n_A}{n} \cdot \frac{n_B}{n} \\
&= \frac{n_A \cdot n_B}{n}
\end{aligned}
$$

Thus, our final estimate, say $\hat{n}$, of the unknown number of errors is:

$$
\hat{n} = \frac{n_A \cdot n_B}{n_{AB}}
$$

Given that together Peter and Paula have detected $n_A + n_B - n_{AB}$ different errors, the number of *remaining* (as yet undetected) errors is estimated to be $\hat{n} - n_A - n_B + n_{AB}$, which after slight algebraic simplification is equal to

$$
\frac{(n_A - n_{AB})(n_B - n_{AB})}{n_{AB}}
$$

For the data given, Peter and Paula had found a total of 25 different errors. The estimate $\hat{n} = \frac{20 \cdot 15}{10} = 30$, so that 5 errors are estimated to have gone undetected by both.

A different approach to this problem is best understood by thinking of Peter and Paula proofreading the text one after another (cf. Feller, 1968, ch. II). Suppose as before that of the total number (which is n) of errors Peter finds exactly n_A, he thus misses $n - n_A$ errors. Next, Paula proofreads the text, and she spots n_B errors. Any of these n_B errors found by her is either one of those n_A errors that had previously also been found by Peter, or it is one of the $n - n_A$ errors that Peter had missed. The probability that among the n_B errors detected by Paula there are n_{AB} of the former and $n_B - n_{AB}$ of the latter type is given by the hypergeometric distribution:

$$
g(n \,|\, n_A, n_B, n_{AB}) = \frac{\binom{n_A}{n_{AB}} \cdot \binom{n - n_A}{n_B - n_{AB}}}{\binom{n}{n_B}}, \qquad \text{where } n \geq n_A + n_B - n_{AB}
$$

---

[1] See PÓLYA, G. (1975).

For given values of n_A, n_B, and n_{AB} this expression depends on the unknown value of n. To estimate it, we maximize g as a function of n. By looking at the ratio $g(n+1)/g(n) \geq 1$ it may be shown that g has two maxima, namely the solution $\hat{n}$ derived earlier, and $\hat{n} - 1$. This is illustrated in Figure 3.11 for the values of $n_A = 20, n_B = 15$, and $n_{AB} = 10$.

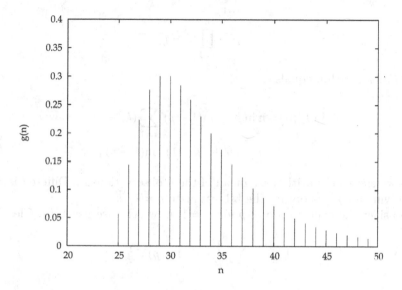

Fig. 3.11. The function $g(n|n_A, n_B, n_{AB})$ for the values of $n_A = 20, n_B = 15$, and $n_{AB} = 10$ has two maxima: one is located at $\hat{n} = 30$, and one at $\hat{n} - 1 = 29$.

3.22 ML Estimation with the Geometric Distribution

a. The likelihood of the data is

$$L(p) = \prod_{i=1}^{n} p \cdot (1-p)^{k_i - 1}$$

$$= p^n \cdot \prod_{i=1}^{n} (1-p)^{k_i - 1}$$

and so its logarithm equals

$$\ln L(p) = n \ln p + \ln(1-p) \sum_{i=1}^{n} (k_i - 1)$$

$$= n \left[\ln p + (\overline{k} - 1) \cdot \ln(1-p) \right]$$

where $\overline{k}$ denotes the arithmetic mean of the n observations k_i. Differentiating and solving for p, we obtain the ML estimate $\hat{p} = 1/\overline{k}$.

Consider the expectation of $\hat{p}$ for $n = 1$, in which case $\hat{p} = 1/k_1$. Clearly,

$$\mathsf{E}[\hat{p}] = \sum_{k=1}^{\infty} \frac{1}{k} \cdot p \cdot (1-p)^{k-1}$$

$$= p + \frac{1}{2} p(1-p) + \cdots$$

$$> p$$

so that $\hat{p}$ will overestimate p (for an explicit result, see Solution 3.40, and Figure 3.28 there).

b. From the preceding expression for $\ln L$ we have

$$-\frac{\partial^2 \ln L}{\partial p^2} = n \left[\frac{1}{p^2} + \frac{\overline{k} - 1}{(1-p)^2} \right]$$

Now, $\mathsf{E}[\overline{k}]$ clearly equals $1/p$, because the expectation of each individual realization k_i equals $1/p$, and so

$$-\mathsf{E}\left[\frac{\partial^2 \ln L}{\partial p^2} \right] = n \left[\frac{1}{p^2} + \frac{\frac{1}{p} - 1}{(1-p)^2} \right]$$

$$= \frac{n}{(1-p)\,p^2}$$

Therefore, the a.s.e.$(\hat{p})$ will be equal to $p\sqrt{(1-p)/n}$. As Figure 3.12 illustrates, this function is maximal when $p = 2/3$, in which case the a.s.e. is equal to $0.385/\sqrt{n}$.

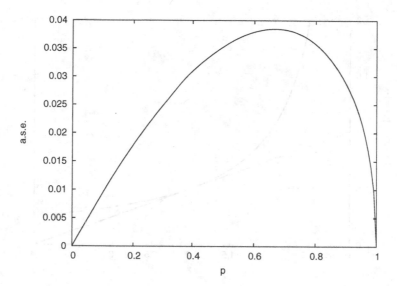

Fig. 3.12. The asymptotic standard error of $\hat{p}$ as a function of the actual success probability, p. The size of the sample from which $\hat{p}$ was obtained is $n = 100$. The $a.s.e.(\hat{p})$ is maximal for $p = 2/3$.

It seems mildly surprising that the $a.s.e.(\hat{p})$ has a maximum in p. One way to look at it is that as $p \to 1$, typically the realizations k_i will all be either 1 or at most 2, say, so that there will not be much variability from sample to sample in the statistic $\hat{p} = 1/\overline{k}$ in this case. On the other hand, when $p \to 0$, the k_i will typically all be quite large (and thus so will be their average, $\overline{k}$), in which case $\hat{p} = 1/k$ will nearly always be a small, positive number that, in absolute terms, will not vary much from sample to sample. Therefore, one would expect some intermediate value of p to have a maximal $a.s.e.(\hat{p})$.

c. Consider the function $f(x) = 1/x$ in the neighborhood of the fixed point $x_0 = 1/p$ (see Figure 3.13, for the value of $p = \frac{1}{2}$). By Taylor's expansion, for x close to x_0 we have ($f'(x) = -1/x^2$)

$$f(x) \approx f(x_0) + (x - x_0)f'(x_0)$$
$$= f\left(\frac{1}{p}\right) + (x - \frac{1}{p})f'(\frac{1}{p})$$
$$= p - (x - \frac{1}{p})p^2$$
$$= 2p - xp^2$$

We already know that $a.s.e.(\hat{p}) \to 0$ as $n \to \infty$. Therefore, when n is large, $\hat{p} = 1/\overline{k}$ must be close to p and $\overline{k}$ be close to $1/p$, so that the preceding linear approximation will be satisfactory. Inserting into the approximation,

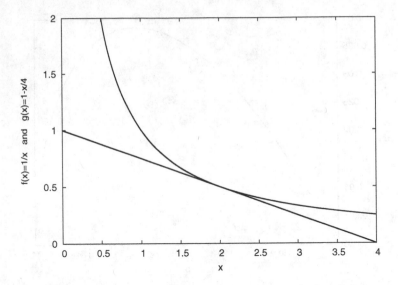

Fig. 3.13. The function $f(x) = 1/x$ and its first-order Taylor expansion $g(x) = 1 - x/4$ around the point $x_0 = 1/p$, here illustrated for $p = \frac{1}{2}$, so that $x_0 = 2$. Clearly, if the sample mean $x = \overline{k}$ is in the ± 0.25 neighborhood of the expectation $x_0 = 1/p$ (which will be the case if n is not too small), then the linear approximation of $g(\overline{k}) = 1 - \overline{k}/4$ to $f(\overline{k}) = 1/\overline{k}$ will be satisfactory.

$$\hat{p} = f(\overline{k}) = \frac{1}{\overline{k}} \approx 2p - \overline{k}\, p^2$$

As a mean, $\overline{k}$ will tend to normality, and so therefore will $\hat{p}$ as a linear approximation to it. For example, because $E[K]$ is equal to $1/p$, it is clear that $E[\hat{p}] = 2p - E[\overline{k}]p^2 = p$ if the linear approximation holds, that is, if n is sufficiently large. Indeed, the approximation gives us an easy alternative way (using the fact given on page 1 that $\text{Var}[K] = (1 - p)/p^2$) to determine the asymptotic variance of $\hat{p}$, not via the Fisher information as earlier, but simply by determining the variance of $2p - \overline{k}\, p^2$.

3.23 How Many Twins Are Homozygotic?

a. From the assumptions stated, the probabilities for any given twin pair to be mm, ff, or mf are

$$p_{mm} = \alpha\beta + (1 - \alpha)\beta^2$$
$$p_{ff} = \alpha(1 - \beta) + (1 - \alpha)(1 - \beta)^2$$
$$p_{mf} = 2(1 - \alpha)\beta(1 - \beta)$$

For example, for twins to be mm, they could either be homozygotic (probability α) and their common sex is male (probability β), or dizygotic (probability $1 - \alpha$) and both of them independently become males (probability $\beta \cdot \beta$). The probabilities p_{ff} and p_{mf} arise in similar ways; clearly, $p_{mm} + p_{ff} + p_{mf} = 1$.

With these probabilities at hand, the likelihood of the data is given as

$$L(\alpha, \beta) = (p_{mm})^{n(mm)} \cdot (p_{ff})^{n(ff)} \cdot (p_{mf})^{n(mf)}$$

where, for example, $n(mm)$ is the number of pairs in which both twins are male. Taking the logarithm of L and maximizing it in α, β yields the ML estimates $\alpha = 0.2872$ and $\beta = 0.5065$ for the data reported by von Bortkiewicz (1920). The shape of the log likelihood function is sketched as a contour plot in Figure 3.14.

It seems quite remarkable that from the observed sex-related twin classification alone, without any invasive genetical analysis, we are able to infer that about 29% of these twin pairs must have been homozygotic. Also, for each genetic sex determination, the outcome "male" was slightly more probable ($\beta = 0.5065$) than "female" ($1 - \beta = 0.4935$), a small but reliable difference of about 1.3% that has long been known.

b. Is this latter effect significant? If we maximize the preceding likelihood $L(\alpha, \beta)$ subject to the constraint that $\beta = \frac{1}{2}$, then the corresponding ML estimate for the probability of a homozygotic twin pair is $\alpha = 0.2873$, nearly as before. Comparing the two models via the likelihood ratio statistic λ described in the Hints, we get $\lambda = 4.70$, $df = 1$, which has a significance level of about 0.03. Thus, the restriction of $\beta = \frac{1}{2}$ has significantly decreased the fit of the model, so that the effect $\beta > \frac{1}{2}$ is probably real (as is strongly suggested by other, independent evidence).

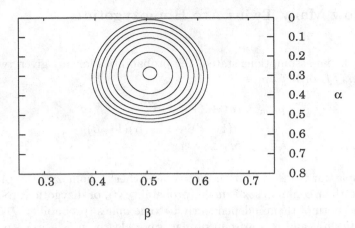

Fig. 3.14. Contour plot of the log likelihood function for the data of von Bortkiewicz (1920); the lines connect points of equal likelihood. The maximum occurs at the point $(\alpha = 0.2872 \,|\, \beta = 0.5065)$. The sketch indicates that for any given value of α the likelihood is maximal near $\beta = \frac{1}{2}$.

3.24 The Lady Tasting Tea

a. Suppose that among the set of n cups the lady designates as type T there are k hits. Thus, the remaining $n - k$ cups of this set must be of type M, and so the remaining k cups of type M will be members of the second set designated as type M (i.e., k hits). This reasoning may be seen more clearly by considering the following table:

		The lady's response:		
		"Tea first"	"Milk first"	Sum
Infusions' true state:	Tea first	k		n
	Milk first			n
	Sum	n	n	$2n$

The margins of the table capture the boundary conditions of the experiment: there are n cups of either type, and the lady has to designate exactly n of the $2n$ cups as being of type T and exactly n as being of type M. Given these boundary conditions, the only way to complete this table is

		The lady's response:		
		"Tea first"	"Milk first"	Sum
Infusions' true state:	Tea first	k	$n - k$	n
	Milk first	$n - k$	k	n
	Sum	n	n	$2n$

b. Our strategy is to first look at how many cups of either type the lady will "directly" identify. Given this information, we next ask for the (conditional) probability that she completes her set of n to-be-selected type T cups in such a way so as to obtain k hits.

Denote as $\mathbf{X}_1, \mathbf{X}_2$ the number of type T and type M cups, respectively, that the lady identifies correctly. Clearly, according to the model the probability of the event $(\mathbf{X}_1 = x_1, \mathbf{X}_2 = x_2)$ is the product of the two binomials

$$\mathsf{P}(\mathbf{X}_1 = x_1, \mathbf{X}_2 = x_2) = \binom{n}{x_1} \theta^{x_1}(1 - \theta)^{n - x_1} \binom{n}{x_2} \theta^{x_2}(1 - \theta)^{n - x_2}$$

Conditional on $\mathbf{X}_1 = x_1, \mathbf{X}_2 = x_2$, $0 \leq x_1, x_2 \leq k$, there remain $n - x_1$ type T and $n - x_2$ type M cups that the lady has not been able to identify, a total of $2n - x_1 - x_2$ cups. Out of this total, she needs to select, on a random guessing basis, further $n - x_1$ cups to complete her set of n cups designated as type T. In order to score k hits among this set, it is then necessary that among the $n - x_1$ further cups she selects there will be $k - x_1$ type T cups and thus $n - k$ type M cups, an event that has the hypergeometric probability

$$\frac{\binom{n-x_1}{k-x_1}\binom{n-x_2}{n-k}}{\binom{2n-x_1-x_2}{n-x_1}}$$

Note that if the lady identifies more than k type T cups or more than k type M cups, then her number of hits will necessarily exceed k. On summing across all contributing values $0 \leq x_1, x_2 \leq k$, and collecting powers to the same base, we find

$$p_n(k|\theta) = \sum_{x_1=0}^{k} \sum_{x_2=0}^{k} \binom{n}{x_1} \binom{n}{x_2} \frac{\binom{n-x_1}{k-x_1}\binom{n-x_2}{n-k}}{\binom{2n-x_1-x_2}{n-x_1}} \cdot \theta^{x_1+x_2} (1 - \theta)^{2n-x_1-x_2}$$

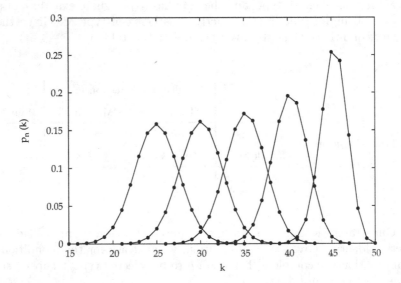

Fig. 3.15. The probability distributions $p_n(k|\theta)$ of hits (k) among $n = 50$ cups of either sort for $\theta = 0.0, 0.2, 0.4, 0.6, 0.8$.

Note that if $\theta = 0$, then $p_n(k|\theta)$ reduces to the term with $x_1 = x_2 = 0$ (i.e., $0^0 = 1$) and thus to the hypergeometric distribution. Figure 3.15 illustrates how the distributions of correctly identified cups of each type (out of $n = 50$

for each type) change with increasing values of the sensitivity parameter, θ. For $\theta = 0$, the lady will on average just guess half ($k = 25$) of all cups of each type correctly, and chance variations around this value are symmetric. On the other hand, with $\theta = 0.8$, the lady will on average classify 45.2 cups of each type correctly, and only rarely ($p = 0.042$) will she fall below $k = 43$. It is equally clear from Figure 3.15 that the lady will face considerable problems to prove her ability beyond reasonable doubt, relative to pure guessing behavior, even with a design using as many as 2×50 cups, if her θ is positive, but smaller than about 0.4.

c. For each n, we first define a critical level, $k_c(n)$, which is the smallest value of k reached or exceeded only with probability $\alpha(n) \leq 0.05$ if the lady is simply guessing (i.e., if $\theta = 0$). To demonstrate her ability with n cups of each type, the lady needs to reach or exceed $k_c(n)$. From the preceding solution with $\theta = 0.50$, the probability $\beta(n)$ of this event is

n	$k_c(n)$	$\alpha(n)$	$\beta(n)$
4	4	0.014	0.302
10	8	0.012	0.571
20	14	0.013	0.881

Thus, with just 4 cups of each type, there is little hope ($\beta(4) = 0.302$) for the lady to demonstrate her ability because she would need to classify all of the cups correctly. Even with 10 cups of each type (in which case 8 or more hits are required), she will still fail to convince the skeptic with probability $1 - \beta(10) = 0.429$. Clearly, then, even with a substantial ability as indexed by $\theta = 0.50$, it needs quite a number of independent trials to grant her a fair chance to demonstrate that indeed she does better than guessing.

3.25 How to Aggregate Significance Levels

a. First note that the p-value observed in a single application of the test is an rv, say $\mathbf{U}$. This simply reflects the fact that even when H_0 is true, on repeating the test with independent data the obtained p-values will vary from replication to replication. What, then, is the distribution of $\mathbf{U}$?

Under H_0 the observed p-value $\mathbf{U}$ is defined by the relation $\mathbf{U} = 1 - F(\Lambda)$, and so

$$
\begin{aligned}
P(\mathbf{U} \geq u) &= P[1 - F(\Lambda) \geq u] \\
&= P[F(\Lambda) \leq 1 - u] \\
&= P[\Lambda \leq F^{-1}(1 - u)] \quad \text{(where } F^{-1} \text{ is the inverse of } F) \\
&= F[F^{-1}(1 - u)] \\
&= 1 - u
\end{aligned}
$$

so that $\mathbf{U}$ is uniform on $[0, 1]$.

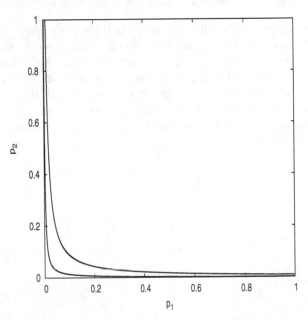

Fig. 3.16. The function $p_2 = k/p_1$ for $k = 0.00865$ and $k = 0.00131$. An aggregate test would be judged overall-significant at the level $\alpha = 0.05$ if the point (p_1, p_2) falls below the hyperbola $p_2 = 0.00865/p_1$. Under H_0, the probability of this event is given by the area $(= \alpha)$ under the hyperbola.

b. At a (too) quick glance one might think that the overall p-value is simply equal to $p_1 \cdot p_2$. However, if this were true, then it would be possible to decrease the aggregate p-value simply by accumulating more and more independent tests, as $p_1 \cdot p_2 \cdot \ldots$ gets smaller with each additional test.

A more convincing procedure is based on the fact observed in a. that (if H_0 is true) $\mathbf{U}$ is uniform, and so $\mathbf{V} = -2 \ln \mathbf{U}$ has an exponential distribution, with mean 2, which is the same as a χ^2-distribution with $df = 2$. Thus, from the additivity of χ^2, when k tests are carried out, the aggregate

$$\sum_{i=1}^{k} \mathbf{V}_i = -2 \sum_{i=1}^{k} \ln \mathbf{U}_i$$

will have a χ^2-distribution with $df = 2k$. In particular, for $k = 2$

$$P(\mathbf{V}_1 + \mathbf{V}_2 > t) = \exp\left(-\frac{t}{2}\right) \cdot \left(1 + \frac{t}{2}\right)$$

which for $t = -2\ln(0.09) - 2\ln(0.07) = 10.134$ yields an aggregate p-value of 0.038. Therefore, when two independent tests are individually not significant at the level of, say, $\alpha = 0.05$, then their aggregate may well be significant at this same level. On the other hand, after obtaining a significant $p_1 < \alpha$ from a first data set, the observed p-value from the second data set may be such that overall the aggregate from (p_1, p_2) is not significant at the level α.

c. A simple numerical computation — or a look into a χ^2-table for $df = 4$ — shows that $(1 + t/2) \cdot \exp(-t/2) = 0.05$ if $t = 9.49$. Thus, to get a significant aggregate p-value at the level of $\alpha = 0.05$, we need

$$-2 \ln \mathbf{U}_1 - 2 \ln \mathbf{U}_2 > 9.49 \text{ or}$$
$$\mathbf{U}_2 < 0.00865/\mathbf{U}_1$$

Therefore, when drawn into a unit square, the set of points (p_1, p_2) that falls below the hyperbola $p_2 = 0.00865/p_1$ would be judged overall-significant at the level $\alpha = 0.05$. Under H_0 the probability of (p_1, p_2) falling below this line would be 0.05 (see Figure 3.16). For $\alpha = 0.01$, the corresponding hyperbola is $p_2 = 0.00131/p_1$.

Finally, we should realize that even though the rule to aggregate $-2 \ln p$ across different tests is simple and general, other aggregation rules exist that can be more powerful. For example, in aggregating two independent t-tests from the same population, it is more powerful to merge the two samples (and not to aggregate their ps) so as to obtain a single overall value of t and from it, in turn, an overall value of p.

3.26 Approximately How Tall Is the Tallest?

a. M_n is smaller than m if and only if all n realizations of U are smaller than m, and so its DF is $[F(m)]^n$. Also note that the rv $F(U)$ is uniformly distributed, which means that $P[F(U) \le z] = z$. From these facts, we obtain, successively

$$P[M_n \le F^{-1}(1 - x/n)] =$$
$$\{P[U \le F^{-1}(1 - x/n)]\}^n =$$
$$\{P[F(U) \le (1 - x/n)]\}^n = (1 - x/n)^n$$

b. First note that the last expression tends to e^{-x} if n is not too small. Next, let $m = F^{-1}(1 - x/n)$; solving for x we get $x = n\,[1 - F(m)]$. Inserting these approximations and relations,

$$G_n(m) = P(M_n \le m) \approx \exp\{-n \cdot [1 - F(m)]\}$$

This approximation is reasonable, even for moderate values such as $n \ge 10$, and is excellent for $n \ge 25$. Therefore, if, as m increases, $1 - F(m)$ tends to a simple limiting form (such as a power or exponential function), then the last equation gives the limiting distribution of M_n as n gets large.

It is important to realize here that for medium or large n, the only region of interest, as far as M_n is concerned, is the rightmost tail of F. To illustrate, the interval $[-0.68, +0.68]$ contains the middle 50% of the standard normal distribution, but the middle 50% of the maximum from $n = 100$ independent standard normal rvs are contained in the interval $[2.20, 2.76]$.

c. For the exponential distribution, $1 - F(m) = \exp(-\lambda m)$, and so

$$G_n(m) \approx \exp\{-n\,e^{-\lambda m}\}$$
$$= \exp\{-e^{-[\lambda m - \ln(n)]}\}$$

Therefore, M_n tends to the double exponential distribution with a location parameter that increases in proportion to $\ln(n)$.

d. The two DFs are sketched in Figure 3.17. Solving from the expression for G_n derived earlier, the medians are for males 216.8 cm ($n = 4$ million males) and 221.7 cm ($n = 120$ million males); the corresponding figures for females are 205.7 cm and 209.9 cm.

The effect of the population size is plausible if one considers the population of 120 million as consisting of 30 subpopulations of 4 million each. It is clear that 30 independent "trials" (each consisting of looking at the maximum of $n = 4$ million) will usually lead to a larger overall maximum than just one such "trial." The mean base height difference of 6 cm between males (176 cm) and females (170 cm) translates into a difference of more than 11 cm between the tallest man and the tallest woman in the smaller population, and into a corresponding difference of nearly 12 cm in the larger population. Looking at

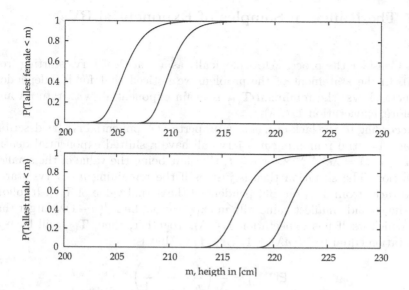

Fig. 3.17. Lower panel: the DF of the height of the tallest male in a population of 4 million (left curve), and 120 million males (right curve). In both populations, height is normally distributed with a mean of 176 cm and a standard deviation of 8 cm. Upper panel: same as bottom, but for females whose height is normally distributed with a mean of 170 cm and a standard deviation of 7 cm.

extremes magnifies and exaggerates the differences seen at the more central representative portions of the distributions. Note also that even the tallest woman in a population of 120 million females will still be about 7 cm shorter than the tallest man from a population of just 4 million males.

3.27 The Range in Samples of Exponential RVs

a. Consider the process chronologically as a "race" of n rvs starting from $t = 0$. In the statement of the problem we noticed that for n independent exponential rvs, the minimum $\mathbf{T}_{(1)}$ is again exponential, with rate $n\lambda$ and, therefore, expectation $1/(n\lambda)$.

According to the lack-of-memory property of exponential rvs (as described on page 14), the remaining $n - 1$ rvs all have a shifted exponential density $k(u|\mathbf{T}_{(1)} = t) = \lambda e^{-\lambda(u-t)}$ for $u \geq t$, the shift being the value of the smallest realization. The situation then is just as if the remaining $n - 1$ rvs start a "fresh race" from $\mathbf{T}_{(1)}$ on, independent of the actual value of $\mathbf{T}_{(1)}$. It follows that the second-smallest value will, in expectation, be $1/[(n-1)\lambda]$ larger than $\mathbf{T}_{(1)}$ which itself has expectation $1/(n\lambda)$. Together, then, $\mathbf{T}_{(2)}$ will have an expectation equal to $1/(n\lambda) + 1/[(n-1)\lambda]$, that is,

$$E[\mathbf{T}_{(2)}] = \frac{1}{\lambda}\left(\frac{1}{n} + \frac{1}{n-1}\right)$$

Continuing this same argument, we see that for $1 \leq k \leq n$

$$E[\mathbf{T}_{(k)}] = \frac{1}{\lambda}\sum_{i=1}^{k}\frac{1}{n+1-i}$$

Now, the expectation of the rv $\mathbf{R}_n$ is

$$\begin{aligned}
E[\mathbf{R}_n] &= E[\mathbf{T}_{(n)} - \mathbf{T}_{(1)}] \\
&= E[\mathbf{T}_{(n)}] - E[\mathbf{T}_{(1)}] \\
&= \frac{1}{\lambda}\sum_{i=1}^{n}\frac{1}{n+1-i} - \frac{1}{n\lambda} \\
&= \frac{1}{\lambda}\sum_{i=2}^{n}\frac{1}{n+1-i} \\
&= \frac{1}{\lambda}\sum_{i=1}^{n-1}\frac{1}{i}
\end{aligned}$$

Clearly, for moderately large n this expression tends to $(\ln n)/\lambda$.

b. Consider again that given some minimum value $\mathbf{T}_{(1)}$ the remaining $n - 1$ rvs are independent shifted exponentials with origin $\mathbf{T}_{(1)}$. Therefore, the probability that they all exceed $\mathbf{T}_{(1)}$ by an amount less than or equal to $r > 0$ is equal to

$$P[\mathbf{R}_n \leq r] = [1 - \exp(-\lambda r)]^{n-1}$$

which is the DF we were looking for.

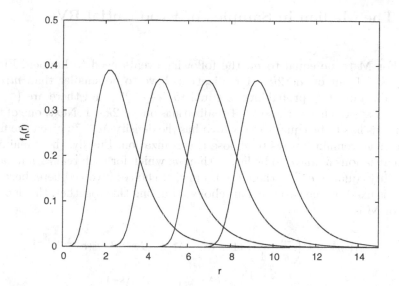

Fig. 3.18. The density $f_n(r)$ of the range of (from left to right) $n = 10, 100, 1000,$ 10000 exponential rvs with rate $\lambda = 1$.

Clearly, for large values of n, this DF will differ only very little from the DF of the maximum $\mathbf{T}_{(n)}$, which is $[1 - \exp(-\lambda r)]^n$. This is in keeping with the fact that for large n, the minimum $\mathbf{T}_{(1)}$ tends to zero, so that effectively the range tends to the maximum.

By differentiation, the corresponding density is

$$f_n(r) = (n - 1)\lambda \exp(-\lambda r)\, [1 - \exp(-\lambda r)]^{n-2}$$

In hindsight, it is possible to "read" the form of this density intuitively as follows. For the range to be equal to r, the following events are needed. First, of those $n - 1$ rvs that are larger than the minimum $\mathbf{T}_{(1)}$, exactly $n - 2$ exceed $\mathbf{T}_{(1)}$ by less than r; this event has probability $[1 - \exp(-\lambda r)]^{n-2}$. Second, one of those $n - 1$ rvs exceeds $\mathbf{T}_{(1)}$ exactly by r, the density of this event is $\lambda \exp(-\lambda r)$. Finally, there are $n - 1$ mutually exclusive ways to pick this latter rv, which together yields the formula for the density f_n.

Four examples ($n = 10, 100, 1000, 10000$) of f_n are graphed in Figure 3.18 that show the basic logarithmic relation that multiplicative (factor of 10) progressions of n translate into additive shifts of f_n of about $\ln 10 \approx 2.30$.

3.28 The Median in Samples of Exponential RVs

a. For **M** to be equal to m, the following events need to happen. First, exactly $k - 1$ out of the $2k - 1$ realizations have to be smaller than m; for each realization, this probability is equal to $1 - e^{-\lambda m}$, and there are $\binom{2k-1}{k-1}$ different ways to choose these $k-1$ realizations out of $2k-1$. Next, one of the realizations has to be equal to m, which has the density $\lambda e^{-\lambda m}$; given the first choice, there remain k ways to choose this realization. Finally, the remaining $k-1$ realizations all have to be larger than m, which for each realization has a probability equal to $e^{-\lambda m}$; there are no further choices involved here, because this set is fixed by the two previous choices. Putting this together, the density g_{2k-1} of **M** is

$$g_{2k-1}(m) = k \binom{2k-1}{k-1} (1 - e^{-\lambda m})^{k-1} \cdot \lambda e^{-\lambda m} \cdot (e^{-\lambda m})^{k-1}$$

$$= \lambda k \binom{2k-1}{k-1} e^{-k\lambda m} \cdot (1 - e^{-\lambda m})^{k-1}$$

b. The direct, but harder way is to multiply $g_{2k-1}(m)$ by m, and to integrate it over the positive reals; in doing so, we need to expand the term $(1 - e^{-\lambda m})^{k-1}$ binomially.

Alternatively, consider that the smallest realization will have an exponential distribution with rate $(2k-1)\lambda$ (see statement (i) on page 14) and thus an expectation equal to the reciprocal of this rate. Next, the distance or "waiting time" from the smallest to the second-smallest realization is again an exponential rv, with rate $(2k - 2)\lambda$. Summing the successive expected distances from one order statistic to the next up to the kth realization in ascending order (i.e., up to the sample median), we directly obtain

$$\mathsf{E}[\mathbf{M}] = \frac{1}{\lambda} \sum_{i=1}^{k} \frac{1}{2k - i} = \frac{1}{\lambda} \sum_{i=k}^{2k-1} \frac{1}{i}$$

Approximating the latter sum by the integral of $1/x$ from k to $2k - 1$, the sum tends, for large k, to $\ln(2k - 1) - \ln(k) = \ln(2 - 1/k) \approx \ln 2$, so that $\mathsf{E}[\mathbf{M}] \to \ln 2/\lambda$, which is the population median, namely the value that each of the generic exponential rvs exceeds with probability of $\frac{1}{2}$. However, for small k, $\mathsf{E}[\mathbf{M}]$ is considerably larger than $\ln 2/\lambda$ (e.g., for $k = 1$, it is equal to $1/\lambda$), and this bias decreases only slowly with k. For example, even when **M** is estimated from a sample of size 15, its still overestimates the true median considerably, in expectation by nearly 5%.

3.29 Breaking the Record

a. If, in the most extreme case, the $n = 100$ realizations form a perfectly increasing order then there will be 100 records. On the other hand, if the first realization already happens to be the largest, then there will be just 1 record. In general, then, the number of records among the first n realizations may be any integer from 1 to n.

The $\mathbf{X}_i$ were assumed to be identically and independently distributed; this implies that all $n!$ permutations of the n realizations are equally likely, only one of which forms a perfectly increasing order. Thus, it is extremely improbable (i.e., $p = 1/n!$) that n records will be observed. Similarly, there are $(n-1)!$ permutations in which the first realization is largest (the other $n-1$ realizations can then be arranged arbitrarily) so that with probability $p = (n-1)!/n! = 1/n$ there will be just 1 record.

Let $\mathsf{E}(n-1)$ be the expected number of records with $n-1$ realizations. Now, if a new, nth realization is added, then it may either exceed the largest value seen in the first $n-1$ realizations, or it may not. In the last case, no new record is added to the previous ones, and in the first case, one further record is added. How likely are these cases? Clearly, the probability that the maximum of n realizations $\mathbf{X}_i$ will occur, specifically, at the last position $i = n$ is just $1/n$ (the maximum may occur at any of the n positions with equal probability). Weighing the two cases by their probabilities,

$$\mathsf{E}(n) = (1 - \frac{1}{n}) \cdot \mathsf{E}(n-1) + \frac{1}{n} \cdot [1 + \mathsf{E}(n-1)]$$

$$= \mathsf{E}(n-1) + \frac{1}{n}$$

Given that $\mathsf{E}(1) = 1$, it follows that

$$\mathsf{E}(n) = \sum_{i=1}^{n} \frac{1}{i}$$

For example, with $n = 100$ realizations there will on average be about 5.19 records. Clearly, $\mathsf{E}(n)$ increases in an approximately logarithmic way with n.

Note that by definition records are determined by ordinal information only. Thus, for any given sequence of realizations, the number of records does not change when all values undergo the same monotone increasing transform.

b. Consider the different, mutually exclusive ways in which the second record takes on the specific value x. First, the rv $\mathbf{X}_1$ must take on some value u, the corresponding density is $f(u)$, where the admissible region is $-\infty < u < x$. For, if already $\mathbf{X}_1 \geq x$ then the second record could not possibly take on the value x. Given that $\mathbf{X}_1 = u$, a variable number $i, i = 0, 1, \ldots$ of further nonrecord realizations may follow that all have to be smaller than or equal to $\mathbf{X}_1 = u$ (otherwise the first of them would be a record). The different values of i correspond to different, mutually exclusive ways whose probabilities add; the probability of i realizations smaller than u equals $[F(u)]^i$. Finally, the next realization needs to attain the value $x > u$; the corresponding density is $f(x)$. Integrating across the admissible region of the variable u we get

$$g_2(x) = \int_{-\infty}^{x} \left[f(u) \cdot [F(u)]^0 \right] du \cdot f(x) + \int_{-\infty}^{x} \left[f(u) \cdot [F(u)]^1 \right] du \cdot f(x) + \dots$$

$$= \int_{-\infty}^{x} \left[f(u) \cdot \left(\sum_{i=0}^{\infty} [F(u)]^i \right) \right] du \cdot f(x)$$

$$= f(x) \cdot \int_{-\infty}^{x} \frac{f(u)}{1 - F(u)} \, du \quad \left(\text{because } \sum_{i=0}^{\infty} z^i = \frac{1}{1-z} \text{ for } 0 \le z < 1 \right)$$

$$= -f(x) \cdot \ln[1 - F(x)]$$

c. We can reason very much as we did in part b. Suppose, then, the claim to hold for some integer r:

$$g_r(x) = f(x) \cdot \frac{1}{(r-1)!} \cdot \{ -\ln[1 - F(x)] \}^{r-1}$$

What, then, is the density $g_{r+1}(x)$? For the $(r+1)$th record to take on the value x, the following needs to be case. First, the rth record must take on some value u, the corresponding density being by assumption $g_r(u)$, where again the admissible region is restricted to $-\infty < u < x$, because otherwise the $(r+1)$th record could not possibly take on the value x. Following the rth record equaling u, a variable number $i, i = 0, 1, \dots$ of nonrecord realizations may arise that, therefore, all have to be smaller than or equal to u. Finally, the next realization needs to take on the value $x > u$, the corresponding density being $f(x)$. Integrating across the admissible region of the variable u we get, using a summation technique as in part b.,

$$g_{r+1}(x) = \int_{-\infty}^{x} \left[g_r(u) \cdot \left(\sum_{i=0}^{\infty} [F(u)]^i \right) \right] du \cdot f(x)$$

$$= f(x) \cdot \int_{-\infty}^{x} \frac{f(u)}{1 - F(u)} \frac{1}{(r-1)!} \, \{ -\ln[1 - F(u)] \}^{r-1} \, du$$

$$= f(x) \cdot \frac{1}{r!} \cdot \{ -\ln[1 - F(x)] \}^{r}$$

which is just the proposition for $g_{r+1}(x)$. Given that the proposition holds for $r = 1$ (when $g_1(x) = f(x)$), it holds generally for any integer r.

Figure 3.19 shows the density $g_r(x)$ of the rth record for $r = 1, 2, 4, 8, 16$ when the parent distribution $f(x)$ (which is shown for $r = 1$) is the standard normal. The means of these distributions necessarily increase; the variances may be seen to decrease with r.

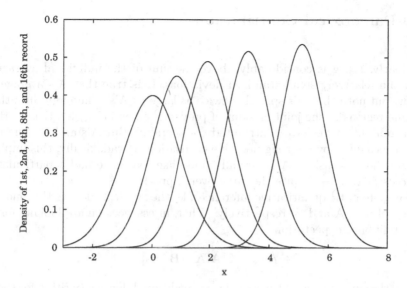

Fig. 3.19. The density $g_r(x)$ of the rth record for (from left to right) $r = 1, 2, 4, 8,$ 16 when the parent distribution ($r = 1$) is standard normal.

d. By the definition of the $(r+1)$th record, it is the first realization following the rth record that is larger than the rth record was. By the lack-of-memory property of exponential rvs (as described on page 14), the $(r + 1)$th record will exceed the rth record by an exponentially distributed excess, whatever the specific value and the distribution of the rth record are. For exponential rvs, the first record is by definition exponentially distributed. The second record exceeds the first record by an exponentially distributed excess; therefore, it is distributed as is the sum of two such exponential rvs. Continuing this argument inductively, we see that the rth record is distributed as is the sum of r independent exponential rvs, each with rate λ. This sum has the gamma (r, λ)-density (cf. page 1)

$$g_r(x) = \lambda e^{-\lambda x} \cdot \frac{(\lambda x)^{r-1}}{(r-1)!}$$

which agrees with our general result for g_r found in part c. For exponential rvs, new records exceed the previous record by an amount that always has the same (exponential) distribution (and therefore, e.g., the same expectation). This is quite different from the case of the normal distribution shown in Figure 3.19, in which old records are exceeded by new ones by an amount that on average gets smaller and smaller as r increases.

3.30 Paradoxical Contribution

Clearly, 175 g is considerably above the sum of the individual averages, given the relatively small standard deviations. It is true that A's average is higher, but note that B's spread is twice as large as A's. Therefore, in rather extreme regions of the joint amount of prey — such as 175 g clearly is — B is much more likely to have contributed considerably. But A's and B's success rates are coupled by a strong negative correlation. Paradoxically, this implies that on those occasions A is actually quite likely to have had a particularly *unsuccessful* day — despite the large overall prey!

For systematic quantitative inferences, let the rvs $\mathbf{A}, \mathbf{B}$ denote the amount of prey of bird A and B, respectively. What, in essence, we are looking for is the conditional expectation

$$\mu_{A|s} = \mathsf{E}[\mathbf{A}|\mathbf{A} + \mathbf{B} = s]$$

that is, the expectation of the associated conditional density (writing for short densities as if they were probabilities)

$$\mathsf{P}(\mathbf{A} = a|\mathbf{A} + \mathbf{B} = s) = \frac{\mathsf{P}(\mathbf{A} = a, \mathbf{B} = s - a)}{\mathsf{P}(\mathbf{A} + \mathbf{B} = s)}$$

The numerator in the last expression is a bivariate normal, and the denominator is a univariate normal, because the sum of two normal rvs is normal again. Inserting the appropriate densities and algebraically simplifying the resulting expression, we obtain again a univariate normal, with an expectation that may be conveniently expressed in multiples of μ_A as follows

$$\frac{\mu_{A|s}}{\mu_A} = \frac{1 + (1 + t)v\varrho + tv^2}{1 + 2v\varrho + v^2}$$

where $v = \sigma_A/\sigma_B$ and $t = (s - \mu_B)/\mu_A$.

The expression gives the conditional mean $\mu_{A|s}$ in units of the unconditional mean μ_A. For the figures given, with $\varrho = -0.8$, we get $\mu_{A|s} = 35$ g, which is actually considerably *less* than A's unconditional average of 60 g. With $\varrho = 0$, we get $\mu_{A|s} = 75$ g, which is still clearly less than B's share of $\mu_{B|s} = 100$ g, despite A's larger overall mean, reflecting again the relatively larger influence of the ratio σ_A/σ_B in extreme regions of the sample space.

The preceding equation permits a quite general analysis of the situation. For example, if $s = \mu_A + \mu_B$, i.e., $t = 1$, then it is generally true that $\mu_{A|s} = \mu_A$, independent of the variances and the correlation. However, even if $\varrho = 0$ the "naive" expectation of a strictly mean-based "proportional share" $\mu_{A|s} = s \cdot \frac{\mu_A}{\mu_A + \mu_B}$ will in general not hold.

3.31 Attracting Mediocrity

a. If σ equals only 3 then it is extremely unlikely that when Peter is tested a result like 105 could obtain — after all, this outcome is five standard deviations above his "true" mean of 90 (see the top panel in Figure 3.20). Therefore, given the measurement of 105 we can be fairly confident that Paula was originally selected. Given that the error in repeated measures is independent and has zero mean, our natural prediction is that the second measurement should be close to 110, Paula's actual IQ.

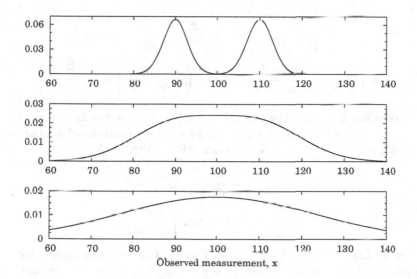

Fig. 3.20. The distribution of observed IQ measurements when Peter (IQ 90) or Paula (IQ 110) is selected at random with $p = \frac{1}{2}$, for different values of σ. For $\sigma = 3$ (top panel), measurements of Peter's and Paula's IQ are clearly separated, so that results such as $x = 105$ will nearly always stem from Paula. However, as σ increases, the large measurement error (middle panel $\sigma = 10$, lower panel: $\sigma = 20$) dilutes the IQ difference between Peter and Paula. Note the different vertical scales for the three panels.

b. Reasoning as in part a., we must realize that with $\sigma = 20$ the measurement error can easily outweigh the base difference of 20 IQ units that exists between Peter and Paula[2], see the bottom panel in Figure 3.20. Specifically, a value like 105 could then nearly as likely stem from a single measurement of Peter's IQ as from one of Paula's. But if either of these cases is approximately

---

[2] Remember that a normal distribution with $\mu = 0$, $\sigma = 20$ effectively ranges from about -40 to $+40$.

equally probable, then our best prediction for the second measurement is the overall mean of Peter's and Paula's IQ, i.e., 100.

More formally, let the indicator rv $\mathbf{I}$ equal 0 if Peter and 1 if Paula was selected. Also, to be somewhat more general, let us assume that Peter's IQ is μ_0, and that of Paula is $\mu_1 > \mu_0$. Bayes' theorem states that the probability that Peter was selected, given the observed measurement x, is

$$
\begin{aligned}
\mathsf{P}(\mathbf{I} = 0 | \mathbf{X} = x) &= \\
&= \frac{\mathsf{P}(\mathbf{X} = x | \mathbf{I} = 0) \cdot \mathsf{P}(\mathbf{I} = 0)}{\mathsf{P}(\mathbf{X} = x | \mathbf{I} = 0) \cdot \mathsf{P}(\mathbf{I} = 0) + \mathsf{P}(\mathbf{X} = x | \mathbf{I} = 1) \cdot \mathsf{P}(\mathbf{I} = 1)} \\
&= \frac{1}{1 + \frac{n(x|\mu_1, \sigma^2)}{n(x|\mu_0, \sigma^2)}} \quad \text{(because } \mathsf{P}(\mathbf{I} = 0) = \mathsf{P}(\mathbf{I} = 1) = \frac{1}{2}) \\
&= \frac{1}{1 + r(x)}
\end{aligned}
$$

$$
\text{where } r(x) = \frac{n(x|\mu_1, \sigma^2)}{n(x|\mu_0, \sigma^2)} = \exp\left\{ \left(\frac{\mu_1 - \mu_0}{\sigma} \right) \left[\frac{x - \frac{\mu_0 + \mu_1}{2}}{\sigma} \right] \right\}
$$

Our prediction $\mathsf{E}_2(x)$ for the second measurement is a weighted mean of μ_0 and μ_1, the weights being the two complementary conditional probabilities $\mathsf{P}(\mathbf{I} = 0 | \mathbf{X} = x)$ and $\mathsf{P}(\mathbf{I} = 1 | \mathbf{X} = x) = 1 - \mathsf{P}(\mathbf{I} = 0 | \mathbf{X} = x)$. That is,

$$
\begin{aligned}
\mathsf{E}_2(x) &= \mu_0 \cdot \frac{1}{1 + r(x)} + \mu_1 \cdot \frac{r(x)}{1 + r(x)} \\
&= \mu_0 + (\mu_1 - \mu_0) \cdot \frac{r(x)}{1 + r(x)}
\end{aligned}
$$

Clearly, if $r(x) \to 1$, then $\mathsf{E}_2(x) \to (\mu_0 + \mu_1)/2$. Our result indicates that this condition will be satisfied when $\sigma \gg \mu_1 - \mu_0$, i.e., when the measurement error dominates the IQ base difference, or when the measurement is close (in terms of σ) to the overall mean, $x \approx (\mu_0 + \mu_1)/2$. In the first case of a large measurement error, any given measurement x may have equally well resulted from selecting Peter or Paula. In the second case of measurements close to the overall mean, the odds also favor neither Peter nor Paula. In both cases it seems reasonable, for lack of more specific evidence, to predict a second measurement that is close to the overall mean, $(\mu_0 + \mu_1)/2$. Finally, if $r(x) \to \infty$, then $\mathsf{E}_2(x) \to \mu_1$. The form of $r(x)$ shows that this condition will be satisfied when either x is exceptionally large, $x \gg (\mu_0 + \mu_1)/2$, or the measurement x is at least above average, $x > (\mu_0 + \mu_1)/2$, and σ is small relative to $\mu_1 - \mu_0$. Both cases strongly suggest that Paula was selected, and so we expect a second measurement that is close to Paula's mean, μ_1.

For the values given in our example, $\mu_0 = 90$, $\mu_1 = 110$, and $x = 105$, we have

$$
r(x) = \exp\left(\frac{100}{\sigma^2} \right)
$$

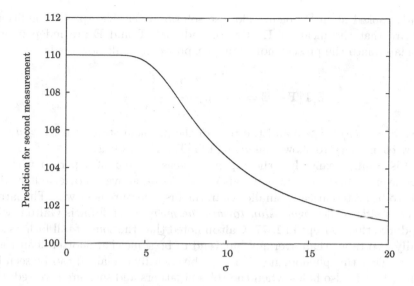

Fig. 3.21. The predicted value of the second IQ measurement as a function of the standard error of the measurement, σ, when the first measurement equaled $x = 105$. Peter's IQ equals 90, Paula's 110.

and so the predicted mean for the second measurement is

$$E_2 = 90 + \frac{20}{1 + \exp\left(-\frac{100}{\sigma^2}\right)}$$

This function is shown in Figure 3.21 which illustrates and summarizes our conclusions reached earlier. Specifically, for $\sigma = 3$, we would expect a value of $E_2 = 109.9997$, very close to Paula's IQ, whereas for $\sigma = 20$, a value of $E_2 = 101.24$ is predicted, quite close to Peter's and Paula's overall IQ mean of 100.

c. Note that a given measurement can be represented as $\mathbf{X} = \mathbf{T} + \mathbf{E}$, the sum of the (randomly selected) true IQ, plus the associated measurement error (which may be positive, zero, or negative). If we knew the true IQ $\mathbf{T}$ of the specific person selected, then we should obviously predict just this value $\mathbf{T}$ as the expected outcome of the second, repeated measurement of that person, because the error of the second measurement has a mean of zero.

Now, we do not actually know the true IQ of the selected person for sure, but we do at least have some indirect evidence. Specifically, the observed value of x attaches to each of the various possible true IQ values a different (conditional) probability. For example, a very high value of x is certainly suggestive of a high true IQ $\mathbf{T}$, especially if σ is not too large. Weighing each possible IQ value by this conditional probability, we get the conditional expectation $E[\mathbf{T}|\mathbf{T} + \mathbf{E} = x]$. This is a special case of a question that we already looked at in Problem 1.30, see solution 3.30. The special conditions here are that the mean of **E** is zero, and that **T** and **E** are independent. Translated into the present notation, our previous result now reads

$$\mathsf{E}[\mathbf{T}|\mathbf{T} + \mathbf{E} = x] = \mu_T \cdot \frac{1 + \left(\frac{x}{\mu_T}\right) \cdot r^2}{1 + r^2}$$

where $r^2 = (\sigma_T/\sigma)^2 > 0$ and the text of the problem stated that $x > \mu_T$. It is now elementary to show that $\mu_T < \mathsf{E}[\mathbf{T}|\mathbf{T} + \mathbf{E} = x] < x$.

This result means that the repeated measurement of a person with an above-average first measurement also tends to be above average, but it also tends to be systematically smaller than the first measurement was. The latter aspect is often called *regression toward the mean*, after Francis Galton, who introduced this concept in 1877. Galton noted that the sons of tall fathers are usually also taller than average, but tend to be somewhat shorter than their fathers. That this phenomenon is not to be causally explained can be seen by the fact that it also holds when the roles of fathers and sons are reversed: the fathers of tall sons are mostly also tall, but not quite as tall as their sons are. Once it is realized, the phenomenon of the regression toward the mean can be recognized in many everyday situations. For example, a restaurant seemed excellent when it was first visited; upon revisiting, it still seems fairly good, but not quite as good as the first time.

An intuitive, informal explanation of this phenomenon is as follows[3]. Consider, to be specific, that **T** is normally distributed with a mean of $\mu_T = 100$ and a standard deviation of $\sigma_T = 10$, as with many IQ tests. Consider that a person was drawn at random, the measured IQ being $x = 130$. What should we expect when the IQ of that same person is measured a second time? Essentially, three groups of people may be distinguished whose measurement led to the observation of $x = 130$. First, there will be people whose true IQ is actually lower than 130 (e.g., 125) but the measurement error happened to be positive (e.g., +5), for example, because they were lucky in guessing the right alternatives. For these people, we would thus predict a value below 130 for the second measurement. Second, there will be people whose true IQ is actually higher than 130 (e.g., 135) but the measurement error happened to be negative (e.g., −5). For these people, we would thus predict a value above 130 for the second measurement. Finally, there are those whose true IQ is actually 130 (or very close to it); for those people, we would predict 130 for the second measurement as well. The crucial point now is that there are *more* people in the first than in the second group: this follows from our assumption that across the population considered, the true IQ has a normal distribution around the mean of $\mu_T = 100$. This assumption implies that there will be more people with an IQ of, say, 125 (first group) than those with 135 (second group). Therefore, we expect more people who on the second measurement

---

[3] see FREEDMAN, D., PISANI, R., AND PURVES, R. (1997).

show a lower rather than a higher result. Complementary considerations show that when the first measurement was below average, the second measurement will still tend to be below average yet somewhat higher, i.e., closer to the mean[4].

---

[4] A general and elegant derivation of a basic related result is due to Das, P. and Mulder, P.G.H. (1983).

3.32 Discrete Variables with Continuous Error

a. Clearly, when $\sigma \to 0$ f will simply tend to $P(\mathbf{N} = n)$, as there will be no significant measurement error. When σ increases to, say, $\sigma = 0.15$, each integer realization of $\mathbf{N} = n$ is distorted by an error that is still small enough to keep the overall measurement $\mathbf{S}$ clearly separated from realizations of $\mathbf{S}$ arising from $\mathbf{N} = n - 1$ or $\mathbf{N} = n + 1$. Thus, each integer will then be surrounded by a component normal density carrying a weight equal to $P(\mathbf{N} = n)$, yielding a density f with multiple separate modes centered around the integers. However, as σ increases even more, say to $\sigma = 0.5$, then repeated measurements $\mathbf{S}$ arising from $\mathbf{N} = n$ and $\mathbf{N} = n + 1$ will show significant overlap. Typically, the resulting density f gets unimodal — if the underlying distribution of $\mathbf{N}$ is itself unimodal — when σ exceeds 0.55 and approaches a smooth, "well-behaved" shape for $\sigma > 0.70$.

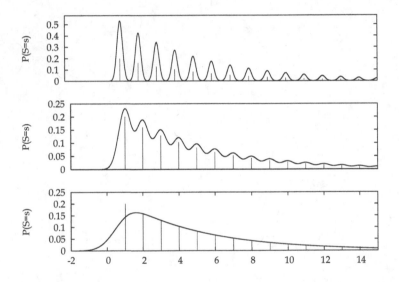

Fig. 3.22. A geometric rv ($p = 0.2$) convolved with normal $N(0, \sigma^2)$ noise. The undistorted distribution of $\mathbf{N}$ is indicated by the discrete vertical lines. Upper panel: $\sigma = 0.15$; middle panel: $\sigma = 0.35$; lower panel: $\sigma = 0.75$. Note the different vertical scale used for the top panel.

Figure 3.22 shows as an example the distibution of a geometric rv that is convolved with noise following a normal $N(0, \sigma^2)$ distribution. If the amount of the noise (σ) is small relative to the unit steps ($n = 1, 2, \ldots$) of the geometric rv (top level), then the discrete nature of the underlying distribution of $\mathbf{N}$ remains clearly visible in the multiple peaks of the resulting distribution of measurements, **S**. However, if σ increases (middle and bottom panels) then the noise dilutes this underlying discrete structure, and the distribution of the measurements assumes a smooth, unimodal form.

b. Let $g(x|\mu, \sigma^2)$ be the normal density with mean μ and variance σ^2. The basic insight is to note that conditional on the event $\mathbf{N} = n$, the observation $\mathbf{S} = s$ if and only if the measurement error — which we assumed follows a normal density with mean 0 and standard deviation σ — is equal to $\mathbf{E} = s - n$. In this way, we get

$$
\begin{aligned}
f(s) &= \sum_{n=0}^{\infty} f(s|\mathbf{N} = n) \cdot \mathrm{P}(\mathbf{N} = n) \\
&= \sum_{n=0}^{\infty} g(s - n|0, \sigma^2) \cdot \mathrm{P}(\mathbf{N} = n) \\
&= \sum_{n=0}^{\infty} g(s|n, \sigma^2) \cdot \mathrm{P}(\mathbf{N} = n)
\end{aligned}
$$

But this is a discrete mixture of normal densities with means $\mu = n$ and common standard deviation σ, the mixture weights being given by $\mathrm{P}(\mathbf{N} = n)$.

3.33 The High-Resolution and the Black-White View

a. The likelihood function for the data as seen by researcher A is

$$L(\mu) = \prod_{i=1}^{n} f(t_i|\mu)$$

$$= \frac{1}{\mu^n} \exp\left(-\frac{1}{\mu} \sum_{i=1}^{n} t_i\right)$$

which implies

$$\ln L(\mu) = -n \cdot \left[\ln(\mu) + \frac{1}{\mu} \cdot \bar{t}\right]$$

where $\bar{t}$ is the arithmetic mean of the t_i. Differentiating this expression gives

$$\frac{\partial \ln L}{\partial \mu} = -n \cdot \left[\frac{1}{\mu} - \frac{1}{\mu^2}\bar{t}\right]$$

$$= \frac{-n}{\mu^2} \cdot [\mu - \bar{t}]$$

Thus, the maximum of the likelihood function occurs at $\hat{\mu} = \bar{t}$, which is the ML estimate obtained by A. As we will reason, $E[\bar{t}] = \mu$, so that this estimate is unbiased.

Also,

$$\frac{\partial^2 \ln L}{\partial \mu^2} = -n \cdot \left[-\frac{1}{\mu^2} + \frac{2}{\mu^3}\bar{t}\right]$$

$$= \frac{n}{\mu^3} \cdot [\mu - 2\bar{t}]$$

Taking expectations, noting that $E[\bar{t}] = \mu$, we get

$$\text{a.s.e.}^2(\hat{\mu}) = \frac{-1}{E\left[\frac{\partial^2 \ln L}{\partial \mu^2}\right]}$$

$$= \frac{\mu^2}{n}$$

Alternatively, because $\hat{\mu}$ is just the sample mean $\bar{t}$, we could have derived this result in the present case, without explicit calculations, from general properties of sums of rvs, as follows. First, we note that the raw data, the exponential rvs t_i, have a mean of μ and variance μ^2. Thus, the sum of the t_i has a distribution (in fact, a gamma, cf. page 1) with mean $n\mu$ and variance $n\mu^2$, which in turn means that the estimate $\bar{t}$ has mean μ and variance μ^2/n.

Thus, in the present case, the s.e. derived from the log-likelihood function holds not only asymptotically, i.e., for large samples, but also for any finite n.

b. From the DF of the t_i, the probability for any single realization to be $\leq c$ equals $1 - \exp(-c/\mu)$, the complementary event has probability $\exp(-c/\mu)$. Thus, for the data as seen by researcher B, the likelihood function is

$$L(\mu) = [\,1 - \exp(-c/\mu)\,]^k \cdot [\,\exp(-c/\mu)\,]^{n-k}$$

or, equivalently

$$\ln L(\mu) = k \ln[\,1 - \exp(-c/\mu)\,] - (n - k)c/\mu$$

Differentiating $\ln L$ with respect to μ gives

$$\frac{\partial \ln L}{\partial \mu} = \frac{c}{\mu^2}\left[n - \frac{k}{1 - \exp(-c/\mu)}\right]$$

On setting this equal to zero and solving for μ, the ML estimate obtained by B is

$$\hat{\mu} = -\frac{c}{\ln(1 - k/n)}$$

This estimate makes sense intuitively: k/n represents an estimate of the probability for a single measurement not to exceed the threshold c; equating this estimate with the corresponding theoretical probability $1-\exp(-c/\mu)$, we obtain the estimate $\hat{\mu}$. A technical problem with this estimate is that it is undefined if $k = 0$ or $k = n$. Therefore we use the modified estimate $\hat{\mu} = -c/[\ln(1-1/2n)]$ (i.e. $k^* = 1/2$) if $k = 0$, and $\hat{\mu} = 0$ if $k = n$.

c. To determine the a.s.e. of B's estimate of μ, we need, as before, the expected second derivative of the log-likelihood. To find this expectation it helps to rewrite the first derivative as follows

$$\frac{\partial \ln L}{\partial \mu} = \frac{c}{\mu^2\,[1 - \exp(-c/\mu)]} \cdot \{n[1 - \exp(-c/\mu)] - k\}$$

Let us rewrite this product formally as $f(\mu) \cdot g(\mu)$. Its derivative is equal to $f'(\mu) \cdot g(\mu) + f(\mu) \cdot g'(\mu)$. Now, the only random quantity — across which we will have to take the expectation — in this expression is k. Its expectation appears only in the first summand, $f'g$, and equals $E[k] = n[1 - \exp(-c/\mu)]$. Therefore, $E[g(\mu)] = 0$, and we are left with only the term $f(\mu) \cdot g'(\mu)$, which is independent of k. In this way we get

$$E\left[\frac{\partial^2 \ln L}{\partial \mu^2}\right] = f(\mu) \cdot g'(\mu) = -\frac{c}{\mu^2\,[1 - \exp(-c/\mu)]} \cdot n \cdot \exp(-c/\mu) \cdot \frac{c}{\mu^2}$$

Simplifying and rearranging, we obtain

$$\text{a.s.e.}^2(\mu) = \frac{-1}{\mathrm{E}\left[\frac{\partial^2 \ln L}{\partial \mu^2}\right]}$$

$$= \frac{\mu^4}{nc^2}[\exp(c/\mu) - 1]$$

The a.s.e. is shown for $\mu = 1$ in Figure 3.23 as a function of the threshold value c.

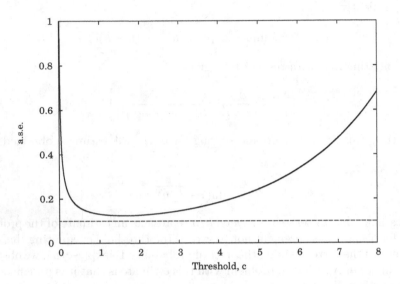

Fig. 3.23. The asymptotic standard error (a.s.e.) of the estimate of the mean from a sample of $n = 100$ exponential rvs with actual mean $\mu = 1$, as a function of the threshold value c used to digitize these rvs. For comparison, the horizontal line shows the a.s.e. of the estimate from the original (nondigitized) rvs, which equals 0.1 in this case. The value of c that minimizes the a.s.e. is 1.59; at this point, the a.s.e. for the estimate from the digitized data is only about 1.24 as large as the a.s.e. of the estimate derived from the original data.

d. Figure 3.23 illustrates that for a given value of μ there is one threshold c that minimizes the a.s.e. of B's estimate. Thus, if one could technically adjust this threshold, one would like to set it such that the resulting estimate of μ is maximally precise. To find this "best" threshold, we need to differentiate a.s.e.$^2(\mu)$ as obtained earlier with respect to c, set this expression to zero, and solve for c. Letting $Q = c/\mu > 0$, this procedure leads to the equation

$$1 - \exp(-Q) = \frac{1}{2}Q$$

which cannot be solved explicitly; however, numerically it is not difficult to compute its root, which equals $Q = 1.59$. This means that

$$c_{opt} = 1.59 \cdot \mu$$

This optimal threshold value seems surprisingly large. Indeed, using it, nearly

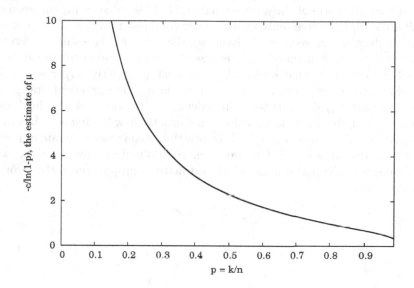

Fig. 3.24. The estimate $\hat{\mu} = -c/[\ln(1 - k/n)]$ as a function of the observed fraction, $p = k/n$, of realizations $\leq c$. Obviously, the larger this ratio, the smaller we will estimate the mean μ to be. The threshold c illustrated here is $1.59 \cdot \mu$, in which case the expected value of p is about 0.80, which would lead to an estimate close to μ. Note that in this region of p small or even medium sampling errors in $p = k/n$ do not influence the estimate $\hat{\mu}$ too much. This is quite different if we choose c smaller so as to decrease the expected fraction (p) of realizations $\leq c$ (consider, e.g., $p = 0.2$, where the curve gets quite steep, so that even relatively small errors in p will strongly influence $\hat{\mu}$).

80% of all realizations will, in expectation, fall below it. Intuitively, one would rather expect a median-like threshold value to be maximally informative — so that about an equal number of realizations would fall above and below it. Also, one might expect threshold values that are exceeded (on average) in 80% versus in 20% to be essentially equivalent, yielding in a sense complementary — and thus equally informative — black-and-white pictures, much like a positive image and its negative. However, the decreasing slope of the function in Figure 3.24 indicates that this intuition neglects that with a relatively high threshold c, the sampling error inherent to the binomial proportion k/n influences the estimate $\hat{\mu}$ much less than with a lower threshold. The value of $c_{opt} = 1.59 \cdot \mu$ represents the best compromise between these conflicting aspects.

How good is B's estimate, relative to that of A? Forming the ratio of the a.s.e. of the estimate obtained by B to that obtained by A, we get

$$\frac{a.s.e. \text{ of B}}{a.s.e. \text{ of A}} = \frac{\mu}{c}\sqrt{\exp(\frac{c}{\mu}) - 1}$$

which depends on the ratio μ/c only. If B uses the optimal value $c_{opt} = 1.59 \cdot \mu$ then the ratio of the a.s.e.s equals 1.24. This value seems surprisingly small: after all, B receives only an extremely condensed version of the original complete data set as seen by A. Even so, the a.s.e. of the estimate derived from these highly condensed data is less than just one-quarter larger than that of A's estimate; that seems like little cost (a slightly larger a.s.e.) for a considerable benefit (much more compact data). One practical drawback, though, is that c_{opt} depends on μ: in order to set the optimal threshold, one would in principle have to know beforehand exactly how large μ is — in which case no estimation would be needed. In practice, some prior estimate of μ will be used in order to select c. The preceding results make it clear that this will usually degrade the performance of B's estimate in proportion to the error of the prior estimate.

3.34 The Bivariate Lognormal

a. The text of the problem provides the basic result that if $\mathbf{U} = \exp(\mathbf{X})$ is a lognormal rv (i.e, $\mathbf{X}$ is a normal rv with mean μ_x and variance σ_x^2), then

$$E[\mathbf{U}] = \exp\left(\mu_x + \frac{1}{2}\sigma_x^2\right)$$

The variance of $\mathbf{U}$ is defined as

$$\text{Var}[\mathbf{U}] = E[(\mathbf{U} - E[\mathbf{U}])^2] = E[\mathbf{U}^2] - E^2[\mathbf{U}]$$

By the definition of the rv $\mathbf{U} = \exp(\mathbf{X})$, we have $\mathbf{U}^2 = [\exp(\mathbf{X})]^2 = \exp(2\mathbf{X})$. Now, the rv $\mathbf{W} = 2\mathbf{X}$ is normal again, it has mean $2\mu_x$ and variance $4\sigma_x^2$. Therefore, the expectation of $\mathbf{U}^2$ can be determined by applying the basic result about $E[\mathbf{U}]$ to the rv $\mathbf{W}$. This gives

$$E[\mathbf{U}^2] = E[\exp(\mathbf{W})]$$
$$= \exp\left(2\mu_x + 2\sigma_x^2\right)$$

where the term $2\sigma_x^2$ results, more explicitly, from $\frac{1}{2}(4\sigma_x^2)$.

On the other hand, again using the result about $E[\mathbf{U}]$ we have

$$E^2[\mathbf{U}] = \left[\exp\left(\mu_x + \frac{1}{2}\sigma_x^2\right)\right]^2 = \exp\left(2\mu_x + \sigma_x^2\right)$$

Inserting these partial results, we finally obtain the variance of $\mathbf{U}$ as

$$\text{Var}[\mathbf{U}] = E[\mathbf{U}^2] - E^2[\mathbf{U}]$$
$$= \exp\left(2\mu_x + \sigma_x^2\right) \cdot \left[\exp\left(\sigma_x^2\right) - 1\right]$$

b. Under the bivariate normal model, $\varrho = 0$ implies that $\mathbf{X}$ and $\mathbf{Y}$ are stochastically independent. Now, if two rvs are independent, then functions of them — such as $\exp(\mathbf{X})$ and $\exp(\mathbf{Y})$ — are independent, too. Because independence implies zero correlation, the correlation between $\mathbf{U}$ and $\mathbf{V}$ is zero as well.

c. If $\varrho = +1$, we may write $\mathbf{Y} = \alpha + \beta\mathbf{X}$ deterministically. From standard regression theory we also know that the regression coefficient $\beta = \varrho\frac{\sigma_y}{\sigma_x} = 1$ under the present assumptions of $\varrho = +1$ and $\sigma_x = \sigma_y$. Thus, $\mathbf{Y} = \alpha + \mathbf{X}$, deterministically. Inserting this relation into the definition of the $\mathbf{U},\mathbf{V}$, we get $\mathbf{U} = \exp(\mathbf{X})$ and

$$\mathbf{V} = \exp(\mathbf{Y}) = \exp(\alpha + \mathbf{X}) = \exp(\alpha) \cdot \exp(\mathbf{X}) = k \cdot \mathbf{U}$$

where $k = e^\alpha$ is a positive constant. Clearly, the correlation of $\mathbf{U}$ with $\mathbf{V} = k \cdot \mathbf{U}$ is perfect.

Why does this reasoning not go through when $\varrho = -1$, even though in the bivariate normal case $\varrho = +1$ and $\varrho = -1$ play perfectly symmetrical roles? The reason is that with $\varrho = -1$ (and still assuming that $\sigma_x = \sigma_y$) the regression coefficient becomes $\beta = -1$, which implies

$$\mathbf{V} = \exp(\mathbf{Y}) = \exp(\alpha - \mathbf{X}) = \frac{\exp(\alpha)}{\exp(\mathbf{X})} = \frac{k}{\mathbf{U}}$$

Whereas $\mathbf{U}$ shows a perfect linear correlation with $k \cdot \mathbf{U}$, that is not true of $\mathbf{U}$ and $k/\mathbf{U}$.

Note that the assumption of $\sigma_x = \sigma_y$ is critical in both cases $\varrho = \pm 1$; if it does not hold, then $\mathbf{U}$ and $\mathbf{V}$ will not be perfectly correlated, even if $\varrho = +1$.

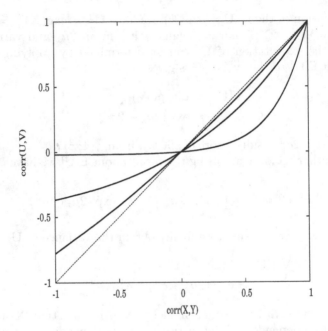

Fig. 3.25. The correlation between $\mathbf{U}$ and $\mathbf{V}$ as a function of the correlation ϱ of the original rvs $\mathbf{X}$ and $\mathbf{Y}$, when $\sigma_x = \sigma_y = \sigma$, for $\sigma = \frac{1}{2}, 1, 2$ (the lowest, middle, top solid line on the left side of the graph, respectively). The comparison with the identity function (the dotted line $y = x$) illustrates that the correlation between $\mathbf{U}$ and $\mathbf{V}$ is always weaker than that between $\mathbf{X}$ and $\mathbf{Y}$, except in the limiting case of $\varrho = +1$. Note that for $\sigma \geq 2$, the correlation is never negative by any substantial amount — even as $\varrho \to -1$!

d. The correlation coefficient is defined as the covariance divided by the product of the standard deviations of the two rvs whose correlation we seek to determine. The covariance in turn is defined as

$$\mathsf{Cov}[\mathbf{U}, \mathbf{V}] = \mathsf{E}[\mathbf{U}\mathbf{V}] - \mathsf{E}[\mathbf{U}] \cdot \mathsf{E}[\mathbf{V}]$$

The last two expectations in this expression are known from the basic result about $E[\mathbf{U}]$ so that the only really new part concerns the expectation of the product $\mathbf{UV}$. Now, by the definition of $\mathbf{U}$, $\mathbf{V}$

$$\mathbf{UV} = \exp(\mathbf{X}) \cdot \exp(\mathbf{Y}) = \exp(\mathbf{X} + \mathbf{Y})$$

Let the rv $\mathbf{S} = \mathbf{X} + \mathbf{Y}$. Clearly, $\mathbf{S}$ is normally distributed, with mean $\mu_s = \mu_x + \mu_y$ and variance $\sigma_s^2 = \sigma_x^2 + \sigma_y^2 + 2\varrho\sigma_x\sigma_y$. It then follows from the basic result about $E[\mathbf{U}]$ that

$$E[\mathbf{UV}] = E[\exp(\mathbf{S})]$$
$$= \exp\left(\mu_s + \frac{1}{2}\sigma_s^2\right)$$

Putting together this result — with μ_s, σ_s expressed in terms of the original parameters as earlier — and the previously known facts about the means and variances of $\mathbf{U}$, $\mathbf{V}$, the correlation coefficient is, after slight simplification, found to be

$$\operatorname{corr}(\mathbf{U}, \mathbf{V}) = \frac{\exp(\varrho\sigma_x\sigma_y) - 1}{\sqrt{\left[\exp(\sigma_x^2) - 1\right]\left[\exp(\sigma_y^2) - 1\right]}}$$

Note that the correlation between $\mathbf{U}$ and $\mathbf{V}$ is completely independent of the means of $\mathbf{X}$ and $\mathbf{Y}$. As explained earlier, the exponentiation turns the *location* parameters μ_x, μ_y of $\mathbf{X}$ and $\mathbf{Y}$ into *scaling factors* of $\mathbf{U}$ and $\mathbf{V}$. Because variations of the scaling factors generally do not change linear correlations, the result was to be expected. On the other hand, the standard deviations σ_x, σ_y of $\mathbf{X}$ and $\mathbf{Y}$ turn into *powers* for $\mathbf{U}$ and $\mathbf{V}$, which generally do influence the linear correlation coefficient.

The result about the correlation of $\mathbf{U}$ and $\mathbf{V}$ may also be used to verify the properties we already discussed in part c. Note, in particular, that in the equal-variance case $\sigma_x = \sigma_y = \sigma$ we get

$$\operatorname{corr}(\mathbf{U}, \mathbf{V}) = \frac{\exp(\varrho\sigma^2) - 1}{\exp(\sigma^2) - 1}$$

The correlation of $\mathbf{U}$ and $\mathbf{V}$ is shown in Figure 3.25 as a function of the correlation ϱ of the original rvs $\mathbf{X}$ and $\mathbf{Y}$, for three values of σ. For $\varrho = +1$, the two correlations are identical, but for any other value the correlation between $\mathbf{U}$ and $\mathbf{V}$ is weaker than that between $\mathbf{X}$ and $\mathbf{Y}$.

3.35 The arcsin($\sqrt{p}$) Transform

a. If n is large, then the estimate $\hat{p}$ will tend to be close to p. That is, the sampling error $\Delta p = \hat{p} - p$ is bound to be small, so that Taylor's expansion holds in a neighborhood of the order of Δp around p. Therefore,

$$g(\hat{p}) = g(p + \Delta p) \approx g(p) + \Delta p \cdot g'(p)$$

The rv $\Delta p = \hat{p} - p$ has $\mathsf{E}[\Delta p] = 0$ and variance

$$\mathsf{Var}[\Delta p] = \mathsf{Var}[\hat{p} - p] = \mathsf{Var}[\hat{p}] = p(1 - p)/n$$

Thus, the expectation of the transformed estimate $g(\hat{p})$ is, approximately,

$$\mathsf{E}[g(\hat{p})] \approx \mathsf{E}[g(p) + \Delta p \cdot g'(p)]$$
$$= g(p) + \mathsf{E}[\Delta p] \cdot g'(p)$$
$$= g(p)$$

and its variance is

$$\mathsf{Var}[g(\hat{p})] \approx \mathsf{Var}[g(p) + \Delta p \cdot g'(p)]$$
$$= \mathsf{Var}[\Delta p] \cdot [g'(p)]^2$$
$$= \frac{p(1 - p)}{n} \cdot [g'(p)]^2$$

b. The last result in part a. indicates that if g would satisfy the differential equation

$$g'(p) = \frac{1}{2\sqrt{p(1 - p)}}$$

then the variance of the transformed estimate, $\mathsf{Var}[g(\hat{p})]$, would indeed be independent of p (the factor 2 in the denominator here is arbitrary; it just helps to avoid trivial scaling issues). Standard analysis techniques[5] show that the solution to this differential equation is

$$g(p) = \arcsin(\sqrt{p})$$

Thus, for this particular choice of g, the variance of the transformed estimate equals $\mathsf{Var}[g(\hat{p})] = 1/(4n)$, which is independent of p.

Figure 3.26 illustrates the principal behavior of the transform g. If the scaled transform $g^*(p) = \frac{2}{\pi} \arcsin(\sqrt{p})$ is used instead, then the transformed p-values fall into the interval $[0, 1]$, just like the original probabilities do.

---

[5] It is simplest to start from the inverse of g: $p = g^{-1}(v) = \sin^2(v)$; next, differentiate p with respect to v to get $dp/dv = 2\sin(v)\cos(v) = 2\sqrt{p(1 - p)}$, which yields the desired result by the differentiation rule for inverse functions.

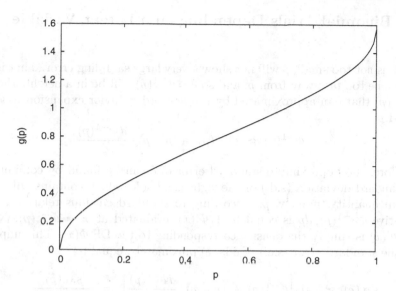

Fig. 3.26. The function $g(p) = \arcsin(\sqrt{p})$ is nearly linear across most of its range. For very small or large p, the curvature of g compensates for the quadratic decrease in the variance of $\hat{p}$ in those regions.

3.36 Binomial Trials Depending on a Latent Variable

If n is not too small, $\hat{p}$ will not show a very large sampling error. That is, $\hat{p}$ will not be too far away from p, and so $\hat{c} = \Phi^{-1}(\hat{p})$ will be in a neighborhood of $\Phi^{-1}(p)$ that can approximated by the first-order Taylor expansion to Φ^{-1} around p:

$$\hat{c} = \Phi^{-1}(\hat{p}) \approx \Phi^{-1}(p) + (\hat{p}-p) \cdot \frac{d\Phi^{-1}(p)}{dp}$$

Therefore, the approximate standard error of $\hat{c}$ can be found by computing the standard deviation (s.d.) of the right-hand side, which contains only one random quantity, namely, $\hat{p}$. According to standard calculus relations, the derivative $d\Phi^{-1}(p)/dp$ is equal to $1/\Phi'(x)$, evaluated at $x = \Phi^{-1}(p)$. Note that $\Phi'(x)$ is simply the density corresponding to the DF $\Phi(x)$. This implies that the standard error (s.e.) of $\hat{c}$ is, approximately, equal to

$$\text{s.e.}(\hat{c}) \approx \text{s.d.} \left[\Phi^{-1}(p) + (\hat{p}-p) \cdot \frac{d\Phi^{-1}(p)}{dp} \right] = \frac{\text{s.d.}(\hat{p})}{\Phi'[\Phi^{-1}(p)]}$$

By standard binomial theory, s.d.$(\hat{p})$ is equal to $\sqrt{p(1-p)/n}$. Estimating p by $\hat{p}$, we find the approximate standard error of $\hat{c}$ as

$$\text{s.e.}(\hat{c}) \approx \frac{\sqrt{\hat{p}(1-\hat{p})/n}}{\Phi'[\Phi^{-1}(\hat{p})]}$$

In the logistic case, considerable simplifications of this somewhat unwieldy expression arise, which probably helped to make this model so popular. From $\Phi(x) = 1/[1 + \exp(-x)]$, we have $\Phi'(x) = \exp(x)/[1 + \exp(x)]^2$; also, the inverse DF is given by $\Phi^{-1}(p) = \ln[p/(1-p)]$. Taken together, this implies that $\Phi'[\Phi^{-1}(p)] = p(1-p)$, and so the approximate standard error of $\hat{c}$ takes the particularly simple form

$$\frac{1}{\sqrt{n\,\hat{p}(1-\hat{p})}}$$

In the normal case, $\Phi'(x)$ is the standard normal density function, and $\Phi^{-1}(\hat{p})$ is the usual "$z_{\hat{p}}$-value" corresponding to $\hat{p}$. In this case, the approximate standard error of $\hat{c}$ is estimated by

$$\exp\left(\frac{1}{2} z_{\hat{p}}^2 \right) \cdot \sqrt{2\pi \hat{p}(1-\hat{p})/n}$$

These two standard errors are compared graphically in Figure 3.27 for $n = 100$. This comparison reveals that the s.e. of $\hat{c}$ is minimal if $p = \frac{1}{2}$. Note that in contrast the s.d. $\sqrt{p(1-p)/n}$ of the associated binomial estimate $\hat{p}$ — the relative frequency of successes seen in the n trials — is maximal, not minimal,

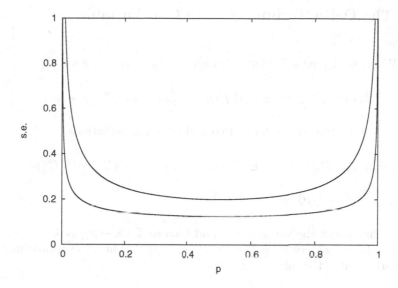

Fig. 3.27. The approximate standard error of $\hat{c}$ under the logistic (upper curve) and the normal model (lower curve) when the estimate is based on $n = 100$ trials.

if $p = \frac{1}{2}$, and this binomial s.d. forms the numerator of s.e.$(\hat{c})$. However, the denominator is also maximal for $p = \frac{1}{2}$, and it exerts the dominating influence on the behavior of the ratio. This fact essentially reflects the geometry of the function $c = \Phi^{-1}(p)$. Near $p = \frac{1}{2}$, small or even medium sampling errors in estimating p will just lead to minor associated errors in the estimate of c because the function $c - \Phi^{-1}(p)$ is relatively flat in this region. This, however, is quite different when p is close to 0 or 1, in which case even small sampling errors $|\hat{p} - p|$ will lead to relatively large errors $|\hat{c} - c|$ because $c = \Phi^{-1}(p)$ is quite steep in those regions.

Figure 3.27 also indicates that under the logistic model the s.d. of $\hat{c}$ is generally larger. Partly, this is simply due to the spread of the logistic distribution, which is larger by a factor of $\pi/\sqrt{3} \approx 1.81$ than for the normal. A second factor relevant here is that with the logistic the approach $p \to 0$ and $p \to 1$ is considerably slower than for the normal. Even if we equate the spread of the logistic model with that of the normal version by the appropriate scale factor, the s.e. for the logistic model is still larger for values of $p < 0.16$ or $p > 0.84$, but it is smaller in between these values.

3.37 The Delta Technique with One Variable

a. We expand f as a Taylor series around $x_0 = \mu$, as follows

$$f(x) = f(\mu) + (x - \mu) f'(\mu) + \frac{1}{2}(x - \mu)^2 f''(\mu) + \cdots$$

Keeping terms up to the order of two and taking expectations,

$$E[f(\mathbf{X})] \approx E[f(\mu)] + E[(\mathbf{X} - \mu)] f'(\mu) + \frac{1}{2}E[(\mathbf{X} - \mu)^2] f''(\mu)$$

$$= f(\mu) + \frac{1}{2}\sigma^2 f''(\mu)$$

by the definition of the variance σ^2, and because $E[(\mathbf{X} - \mu)] = 0$.

b. For $f(x) = e^x$, we have $f'(x) = f''(x) = f(x)$. Thus, the approximation from part a. takes the form

$$E[\exp(\mathbf{X})] \approx e^\mu + \frac{1}{2}\sigma^2 e^\mu$$

$$= e^\mu \left(1 + \frac{1}{2}\sigma^2\right)$$

In this relatively simple case, we already know the exact result from Problem 1.34

$$E[\exp(\mathbf{X})] = \exp\left(\mu + \frac{1}{2}\sigma^2\right)$$

$$= \exp(\mu) \cdot \exp\left(\frac{1}{2}\sigma^2\right)$$

$$= \exp(\mu) \cdot \left(1 + \frac{1}{2}\sigma^2 + \cdots\right)$$

Thus, when σ is small so that $e^{\sigma^2/2}$ can be approximated by the first two terms of the exponential series, $e^{\sigma^2/2} \approx 1 + \frac{1}{2}\sigma^2$, the approximation is acceptable. In practice this requires that $\sigma \leq 1$ in the present example.

More generally, note that the approach in part a. is based on approximating f by a quadratic function around μ. When f is a polynomial up to order two, this approximation is exact; for some functions it is quite acceptable over a considerable range, while still other functions (e.g., $f(x) = e^x$) may only very locally be approximated by a quadratic. The condition for the approximation to work is that the density of $\mathbf{X}$ be concentrated within that region around μ where the quadratic approximates f well. Typically, the approximation is invoked in situations when f has a much more complicated form than e^x so that exact results are hard to obtain (see part c.), and when σ is known to be small — for example because it decreases with the sample size (e.g., when **X** is a sample mean).

c. From elementary mechanics, for a specific angle α (measured in radians[6]), the maximum height (measured in meters, $[m]$) is equal to

$$f(\alpha) \ = \ \frac{v_0^2}{2g} \sin^2(\alpha)$$

where $g = 9.81 \, [m/s^2]$ is the gravitational constant. Therefore, we are looking for $\mathsf{E}[f(\alpha)]$ when the angle α is considered as an rv. In the present case, we find $f''(\alpha) = (v_0^2/g) \cdot \cos(2\alpha)$, and on inserting into our result from part a., we get

$$\mathsf{E}[f(\alpha)] \approx f(\mu_\alpha) \ + \ \frac{1}{2}\sigma^2 \, f''(\mu_\alpha) \ = \ \frac{v_0^2}{2g} \left[\sin^2(\mu_\alpha) + \sigma_\alpha^2 \, \cos(2\mu_\alpha) \right]$$

The problem states that the initial velocity is $v_0 = 120 \, [m/s]$, and that the ball is on average thrown vertically. That is, we have $\mu_\alpha = \pi/2$ (i.e., 90°). Also, the standard deviation measuring the variation of the angle from throw to throw is given as $\sigma_\alpha = 0.1745$ (i.e., 10°). Note that any deflection from the vertical decreases the maximum height the ball will reach. The "naive" prediction — when the variability of the angle α is neglected altogether — of $f(\mu_\alpha) = 733.9 \, [m]$ will therefore necessarily overestimate the true mean. The above series-approximation to $\mathsf{E}[f(\alpha)]$ predicts an expected maximum height of 711.6 $[m]$ – that is, more than 22 $[m]$ lower than the naive prediction. This approximation is in fact quite close to the exact expectation of 712.3 $[m]$.

---

[6] The radian corresponding to an angle α measured in degrees is $2\pi \cdot (\alpha/360°)$.

3.38 The Delta Technique with Two Variables

a. We expand f as a Taylor series around (μ_x, μ_y). To do so, let us write as $f_x(a, b)$ the first partial derivative of f with respect to x, evaluated at the point $(x = a, y = b)$. Similarly, $f_{xy}(a, b)$ is the mixed partial derivative $\frac{\partial^2 f}{\partial x \, \partial y}$ evaluated at $(x = a, y = b)$. With this notation,

$$f(x, y) = f(\mu_x, \mu_y) + (x - \mu_x) f_x(\mu_x, \mu_y) + (y - \mu_y) f_y(\mu_x, \mu_y) +$$
$$\frac{1}{2}(x - \mu_x)^2 f_{xx}(\mu_x, \mu_y) + \frac{1}{2}(y - \mu_y)^2 f_{yy}(\mu_x, \mu_y) +$$
$$(x - \mu_x)(y - \mu_y) f_{xy}(\mu_x, \mu_y) + \cdots$$

Keeping terms up to the order of two and taking expectations,

$$\mathsf{E}[f(\mathbf{X}, \mathbf{Y})] \approx \mathsf{E}[f(\mu_x, \mu_y)] + \mathsf{E}[(\mathbf{X} - \mu_x)] f_x(\mu_x, \mu_y) + \mathsf{E}[(\mathbf{Y} - \mu_y)] f_y(\mu_x, \mu_y) +$$
$$\frac{1}{2} \mathsf{E}[(\mathbf{X} - \mu_x)^2] f_{xx}(\mu_x, \mu_y) + \frac{1}{2} \mathsf{E}[(\mathbf{Y} - \mu_y)^2] f_{yy}(\mu_x, \mu_y) +$$
$$\mathsf{E}[(\mathbf{X} - \mu_x)(\mathbf{Y} - \mu_y)] f_{xy}(\mu_x, \mu_y)$$
$$= f(\mu_x, \mu_y) + \frac{1}{2} \sigma_x^2 f_{xx}(\mu_x, \mu_y) + \frac{1}{2} \sigma_y^2 f_{yy}(\mu_x, \mu_y) + \varrho \sigma_x \sigma_y f_{xy}(\mu_x, \mu_y)$$

where we have used the facts that $\mathsf{E}[(\mathbf{X} - \mu_x)(\mathbf{Y} - \mu_y)] = \varrho \sigma_x \sigma_y$, and that the two expectations involving the first derivatives are zero.

b. For $f(x, y) = x/y$ and $\varrho = 0$, the required partial derivatives are

$$f_{xx} = \frac{\partial^2 f}{\partial x^2} = 0 \qquad \text{and} \qquad f_{yy} = \frac{\partial^2 f}{\partial y^2} = \frac{2x}{y^3}$$

Inserting this into our results from a. gives

$$\mathsf{E}\left[\frac{\mathbf{X}}{\mathbf{Y}}\right] \approx \frac{\mu_x}{\mu_y} + \frac{1}{2} \sigma_y^2 \frac{2\mu_x}{\mu_y^3}$$
$$= \frac{\mu_x}{\mu_y} \left[1 + \left(\frac{\sigma_y}{\mu_y}\right)^2\right]$$

c. First note that when two rvs are independent, so are functions of them. Therefore, their expectations may be factored:

$$\mathsf{E}\left[\frac{\mathbf{X}}{\mathbf{Y}}\right] = \mathsf{E}[\mathbf{X}] \cdot \mathsf{E}\left[\frac{1}{\mathbf{Y}}\right] = \mu_x \cdot \mathsf{E}\left[\frac{1}{\mathbf{Y}}\right]$$

The second factor can now be approximated using the results of Problem 1.37 for functions of a single rv. Specifically, with $f(y) = 1/y$ so that $f''(y) = 2/y^3$, we obtain

$$E\left[\frac{1}{\mathbf{Y}}\right] \approx \frac{1}{\mu_y} + \frac{1}{2}\sigma_y^2\frac{2}{\mu_y^3}$$

$$= \frac{1}{\mu_y}\left[1 + \left(\frac{\sigma_y}{\mu_y}\right)^2\right]$$

confirming our previous result. Note, however, that this reduction to the univariate case relies on our factoring of $f(x,y) = g(x) \cdot h(y)$, say; more complicated choices of f that do not admit a similar factorization will generally require the results obtained in part a.

d. In the case of the F-distribution, both the numerator $\mathbf{U}/m$ and the denominator $\mathbf{V}/n$ have an expectation equal to one; the variance of the denominator is $2/n$. Inserting this into the results from part b., we get

$$E[\mathbf{F}_{m,n}] \approx 1 + \frac{2}{n} = \frac{n+2}{n}$$

For $n > 2$, the exact result is $n/(n-2)$. The error of the approximation is $4/(n(n-2))$, and therefore it decreases rapidly in n.

Note that the approximation correctly predicts three qualitative facts:

1. $E[\mathbf{F}_{m,n}]$ is independent of m,
2. $E[\mathbf{F}_{m,n}]$ generally exceeds one, and
3. this excess decreases with n.

e. In addition to random variations of the angle α, now the initial speed varies according to a normal distribution with $\mu_{v_0} = 120$ [m/s] and $\sigma_{v_0} = 10$ [m/s]. We know that for a given initial speed v_0 and angle α, the maximum height is

$$f(v_0, \alpha) = \frac{v_0^2}{2g}\sin^2(\alpha)$$

Writing $v = v_0$ for short, we get

$$E[f(\mathbf{v}, \boldsymbol{\alpha})] \approx f(\mu_v, \mu_\alpha) + \frac{1}{2}\sigma_v^2 f_{vv}(\mu_v, \mu_\alpha) + \frac{1}{2}\sigma_\alpha^2 f_{\alpha\alpha}(\mu_v, \mu_\alpha)$$
$$+ \varrho\sigma_v\sigma_\alpha f_{v\alpha}(\mu_v, \mu_\alpha)$$

From the assumed independent variation of the initial speed and the angle, we have $\varrho = 0$, so that the last summand drops out. Of the remaining terms, the first and third summands had already been calculated in part c. of Problem 1.37; together they are equal to 711.6 [m]. An easy calculation shows that the derivative required for the second summand is

$$f_{vv} = \frac{\partial^2 f}{\partial v^2} = \frac{1}{g}\sin^2(\mu_\alpha)$$

so that the individual contribution from variations of the initial speed is

$$\frac{1}{2} \sigma_v^2 \frac{1}{g} \sin^2(\pi/2) = \frac{1}{2} (10 \,[\text{m/s}])^2 \frac{1}{9.81 \,[\text{m/s}^2]}$$
$$= 5.1 \,[\text{m}]$$

Thus, the variation of the initial speed adds, in expectation, 5.1 [m] to the maximum height of the ball. Together with our previous results from Problem 1.37, the approximation yields an expected maximum height of 716.7 [m], which is quite close to the true expectation of 717.2 [m]. By comparison, the "naive" prediction — when the variation of v_0 or α is simply neglected — is equal to 733.9 [m].

Similar to our considerations in part c., when the angle and the initial speed vary independently we could again have simply factored the expectation

$$\mathsf{E}[f(\mathbf{v}, \boldsymbol{\alpha})] = \frac{1}{2g} \cdot \mathsf{E}[\mathbf{v}^2] \cdot \mathsf{E}[\sin^2(\boldsymbol{\alpha})]$$

and then treat the factors individually. The maximum height predicted by this simpler approach is 716.5 [m], which is nearly as good as the prediction found earlier. As we saw before, this approach is often simpler, but it depends critically on two features: f must be factorizable, and the rvs need to be independent. In the present problem, both conditions are satisfied.

We may also note that variations of the initial speed *increase* the expected maximum height, whereas the previous problem showed that variations of the angle *decrease* it. Can you see why that is?

3.39 How Many Trials Produced a Given Maximum?

a. From Bayes' formula, we find the conditional probability

$$P[\mathbf{N}=1\,|\,\mathbf{M}=0.9] = \frac{P[\mathbf{M}=0.9\,|\,\mathbf{N}=1]\cdot\frac{1}{2}}{P[\mathbf{M}=0.9\,|\,\mathbf{N}=1]\cdot\frac{1}{2} + P[\mathbf{M}=0.9\,|\,\mathbf{N}=2]\cdot\frac{1}{2}}$$

$$= \frac{P[\mathbf{M}=0.9\,|\,\mathbf{N}=1]}{P[\mathbf{M}=0.9\,|\,\mathbf{N}=1] + P[\mathbf{M}=0.9\,|\,\mathbf{N}=2]}$$

Consider the two conditional probabilities in this expression in turn.

Given that $\mathbf{N}=1$, the density of $\mathbf{M}$ is simply the uniform density itself, i.e., equal to $f(x)=1$ for $0 \le x \le 1$.

Turning to the case of $\mathbf{N}=2$, the DF of the maximum of $\mathbf{N}=2$ rvs is $[F(x)]^2$, where F is the DF of the parent rv $\mathbf{X}$. Differentiating, the associated density is $2f(x)F(x)$, where again f stands for the density of $\mathbf{X}$. For a uniform rv $\mathbf{X}$, we have $F(x)=x$ and $f(x)=1$, and so $2f(x)F(x)=2x$.

Evaluating these two expressions at $x=0.9$ we get

$$P[\mathbf{N}=1\,|\,\mathbf{M}=0.9] = \frac{1}{1+1.8} = \frac{5}{14}$$

and thus the complementary probability $P[\mathbf{N}=2\,|\,\mathbf{M}=0.9] = \frac{9}{14}$. Overall then

$$E[\mathbf{N}\,|\,\mathbf{M}=0.9] = 1\cdot\frac{5}{14} + 2\cdot\frac{9}{14} = \frac{23}{14} = 1.643$$

Given that the unconditional expectation of $\mathbf{N}$ already equals 1.5, conditioning on the relatively large value of $\mathbf{M}=0.9$ increases the expectation of $\mathbf{N}$ only to a remarkably small degree. In fact, even in the most extreme case of $\mathbf{M}=1$, the conditional expectation of $\mathbf{N}$ would increase only slightly up to 1.667.

b. Recall (cf. page 2) that the generating function g of a positive rv $\mathbf{N}$ is defined as

$$g(z) = \sum_{n=1}^{\infty} z^n \cdot p(n)$$

where $p(n) = P(\mathbf{N}=n)$. We will need two elementary properties of g later. First, by direct differentiation with respect to z

$$g'(z) = \sum_{n=1}^{\infty} n \cdot z^{n-1} p(n)$$

Also, by a further differentiation

$$g''(z) = \sum_{n=1}^{\infty} n \cdot (n-1) \cdot z^{n-2} p(n)$$

which upon elementary algebra yields the identity

$$\sum_{n=1}^{\infty} n^2 \cdot z^{n-2} p(n) = \frac{1}{z} \cdot g'(z) + g''(z)$$

We now turn to the evaluation of the conditional probability, much as in part a. By Bayes' theorem,

$$P[\mathbf{N}=n \mid \max(\mathbf{X_1},\ldots,\mathbf{X_N})=x] = \frac{P[\max(\mathbf{X_1},\ldots,\mathbf{X_N})=x \mid \mathbf{N}=n] \cdot p(n)}{\sum_{n=1}^{\infty} P[\max(\mathbf{X_1},\ldots,\mathbf{X_N})=x \mid \mathbf{N}=n] \cdot p(n)}$$

Of course, conditional on the event $\{\mathbf{N}=n\}$, we may replace $\max(\mathbf{X_1},\ldots,\mathbf{X_N})$ with $\max(\mathbf{X_1},\ldots,\mathbf{X_n})$, the maximum of n independent realizations of $\mathbf{X}$, which has the DF F^n and thus the density $n \cdot f \cdot F^{n-1}$. Therefore,

$$P[\mathbf{N}=n \mid \max(\mathbf{X_1},\ldots,\mathbf{X_N})=x] = \frac{n\,f(x)\,F^{n-1}(x)\,p(n)}{\sum_{n=1}^{\infty} n\,f(x)\,F^{n-1}(x)\,p(n)}$$

Next, we determine the conditional expectation of $\mathbf{N}$,

$$\begin{aligned}
E[\mathbf{N} \mid \max(\mathbf{X_1},\ldots,\mathbf{X_N})=x] &= \sum_{n=1}^{\infty} n \cdot P[\mathbf{N}=n \mid \max(\mathbf{X_1},\ldots,\mathbf{X_N})=x] \\
&= \frac{\sum_{n=1}^{\infty} n^2\,f(x)\,F^{n-1}(x)\,p(n)}{\sum_{n=1}^{\infty} n\,f(x)\,F^{n-1}(x)\,p(n)} \\
&= F(x) \cdot \frac{\sum_{n=1}^{\infty} n^2\,F^{n-2}(x)\,p(n)}{\sum_{n=1}^{\infty} n\,F^{n-1}(x)\,p(n)}
\end{aligned}$$

Now, the numerator and denominator of the second factor of this latter expression correspond exactly to the two expressions we derived earlier. Upon inserting them, and after very minor algebraic simplifications, we obtain

$$r(x) = E[\mathbf{N} \mid \max(\mathbf{X_1},\ldots,\mathbf{X_N})=x] = 1 + F(x) \cdot \frac{g''[F(x)]}{g'[F(x)]}$$

c. We only need to insert the partial results for geometric rvs. The generating function of a geometric rv is $g(z) = pz/[1 - z(1-p)]$. Differentiating twice, the ratio

$$\frac{g''(z)}{g'(z)} = \frac{2(1-p)}{1 - z(1-p)}$$

and so, after slight algebraic simplifications

$$r(x) = E[\mathbf{N} \mid \max(\mathbf{X_1},\ldots,\mathbf{X_N})=x] = \frac{1 + (1-p)F(x)}{1 - (1-p)F(x)}$$

It is quite remarkable that, for any choice of $F(x)$, even an exceptionally large value x of the maximum — so that $F(x) \to 1$ — only leads to a conditional expectation for **N** equal to $2/p - 1$, which represents less than a doubling of **N**'s unconditional expectation, $1/p$. For example, if $p = \frac{1}{2}$, then the unconditional expectation $\mathsf{E}[\mathbf{N}] = 2$. Observing an exceptionally large value x as the random maximum (with $F(x) \approx 1$) would then still only yield a conditional expectation for **N** of 3. In this sense, $\mathsf{E}[\mathbf{N}]$ varies only weakly with the value of the random maximum in the geometric case.

d. If **N** can take on only the two values 1 and k (with equal probability) then the generating function is clearly $g(z) = (z + z^k)/2$. Differentiating twice, the ratio

$$\frac{g''(z)}{g'(z)} = \frac{k(k-1) \cdot z^{k-2}}{1 + k \cdot z^{k-1}}$$

and so, after slight algebraic simplifications

$$r(x) \;=\; \mathsf{E}[\,\mathbf{N}|\max(\mathbf{X}_1, \dots, \mathbf{X_N}) = x\,] = \frac{1 + k^2 [F(x)]^{k-1}}{1 + k [F(x)]^{k-1}}$$

Trivially, for $k = 1$, the conditional expectation is generally equal to 1. For $k = 2$, the conditional expectation equals $[1 + 4F(x)]/[1 + 2F(x)]$, as derived in part a. for a uniform parent. However, when $k \gg 1$, observing an exceptionally large value x of the random maximum — so that $F(x) \approx 1$ — leads to a conditional expectation for **N** that is close to $(1 + k^2)/(1 + k) \approx k$. This is in keeping with the fact that very large values of the random maximum increase the odds for **N** being equal to k, not to 1.

3.40 Waiting for Success

a. Clearly, N_1 has the geometric distribution $p(n) = p(1-p)^{n-1}$, where $n = 1, 2, \ldots$, and so its expectation is $1/p$, suggesting the moment estimator $\hat{p} = 1/N_1$. The same estimate is derived from maximizing the likelihood function (see Problem 1.22).

The Taylor series expansion of $-\ln(1-x)$, valid for $-1 \le x < 1$, is

$$\sum_{n=1}^{\infty} \frac{1}{n} x^n = -\ln(1-x)$$

Applying this sum to the present problem, the expectation of $\hat{p}$ is seen to be

$$\mathsf{E}[\hat{p}] = \sum_{n=1}^{\infty} \frac{1}{n} p(1-p)^{n-1} = \frac{p}{1-p} \sum_{n=1}^{\infty} \frac{1}{n}(1-p)^n = -\frac{p}{1-p} \ln p = f(p)$$

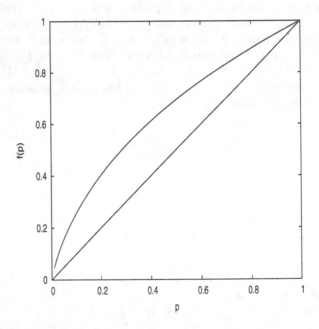

Fig. 3.28. The expectation of the estimate $\hat{p}$ as a function, $f(p)$, of the actual value of p, which is for comparison indicated by the identity function. Clearly, $\mathsf{E}[\hat{p}]$ has a systematic and quite large bias.

Plotting this function as in Figure 3.28 shows that generally $f(p) > p$. For example, when $p = \frac{1}{2}$, the mean estimate $\hat{p}$ is $f(0.5) = 0.693$. Thus, $\hat{p}$ has little to recommend itself.

b. An obvious way to improve our estimate is to look at how long (i.e., how many trials) it takes to get $r > 1$ successes. The expectation for the waiting time to the rth success equals r times the expected waiting time for the first success, because the situation after the first success (now waiting for the second success) is just the same as at the beginning. That is, $E[\mathbf{N}_r] = r \cdot E[\mathbf{N}_1] = r/p$. Thus, if we equate the observed waiting time $\mathbf{N}_r$ with its expectation and solving for p, our moment estimator of the success probability becomes $\hat{p} = r/\mathbf{N}_r$.

How large is the bias of this estimate for $r > 1$? To answer this question we first note that $\hat{p}$ is the reciprocal of the mean of the r individual geometric waiting times from one success to the next. Thus, if r is not too small, $\mathbf{N}_r/r$ will converge to a normal distribution, with mean $\mu - 1/p$, and a variance $\sigma^2 = (1-p)/(r \cdot p^2)$ that decreases in proportion to r.

We can now apply our result from the solution of Problem 1.37 to the rv $\mathbf{X} = \mathbf{N}_r/r$ with $f(x) = 1/x$, so that $f''(x) = 2/x^3$. If the variance of $\mathbf{X} = \mathbf{N}_r/r$ is small (i.e., if r is not too small) then

$$E[r/\mathbf{N}_r] \; = \; E[1/\mathbf{X}] \approx f(\mu) + \frac{1}{2}\sigma^2 f''(\mu)$$

and inserting $\mu = \frac{1}{p}$ we get

$$E[r/\mathbf{N}_r] \approx p + \frac{1}{2} \cdot \frac{1-p}{r \cdot p^2} \cdot 2p^3$$

$$= p + \frac{p(1-p)}{r}$$

Obviously, $\hat{p} = r/\mathbf{N}_r$ still overestimates p, but this bias diminishes as r increases. Actually, if r is not very small, then the approximation is quite good. For example, even if $p = \frac{1}{2}$ and $r = 2$, which is the worst case after $r = 1$, our approximation yields $E[\hat{p}] \approx 5/8 = 0.625$, whereas the true expectation may be shown to be $E[\hat{p}] = 0.614$. Thus, our approximation to $E[\hat{p}]$ is fairly accurate, but for a value as small as $r = 2$, $\hat{p}$ still overestimates p markedly, even though this bias is already considerably smaller than for $r = 1$.

c. For $r > 1$, Haldane (1945) proposed the modified probability estimate $\hat{p} = (r - 1)/(\mathbf{N}_r - 1)$. To find its expectation, we have to weigh $\hat{p}$ by the probabilities for the various possible values of the rv $\mathbf{N}_r$, which has the so-called negative-binomial distribution,

$$p_r(n) \; = \; \binom{n-1}{r-1} p^r (1-p)^{n-r} \;,\; \text{for } n \geq r$$

Thus

$$E[\hat{p}] = \sum_{n=r}^{\infty} \frac{r-1}{n-1} \binom{n-1}{r-1} p^r (1-p)^{n-r}$$

Elementary algebraic operations on this sum will show that it is equal to p. Thus, Haldane's estimate is actually unbiased.

References

BLOM, G., HOLST, L., AND SANDELL, D. (1994). *Problems and Snapshots from the World of Probability*. Springer, New York.

BORTKIEWICZ, L. VON (1920). Die Dispersion der Knabenquote bei Zwillingsgeburten. *Zeitschrift für schweizerische Statistik und Volkswirtschaft*, **56**, 235–246.

BULMER, M.G. (1979). *Principles of Statistics*. (2nd ed.) Dover, New York.

CACOULLOS, T. (1989). *Exercises in Probability*. Springer, New York.

CAPINSKI, M., AND ZASTAWNIAK, T.J. (2001). *Probability through Problems*. Springer, New York.

COX, D.R., AND MILLER, H.D. (1965). *The Theory of Stochastic Processes*. Chapman and Hall, London.

DAS, P., AND MULDER, P.G.H. (1983). Regression to the Mode. *Statistica Neerlandica*, **37**, 15–20.

DERENZO, S.E. (1977). Approximations for hand calculators using small integer coefficients. *Mathematics of Computation*, **31**, 214–225.

FALK, R. (1993). *Understanding Probability and Statistics. A Book of Problems*. Peters, Wellesley.

FELLER, W. (1968, 1971). *An Introduction to Probability Theory and its Applications*. (Vol. I, 3rd ed.; Vol. II, 2nd ed.). Wiley, New York.

FISHER, R.A. (1935). *The Design of Experiments* (1st ed.). Olivier and Boyd, Edinburgh.

FREEDMAN, D., PISANI, R., AND PURVES, R. (1997). *Statistics*. (3rd ed.). Norton, New York.

GRIMMETT, G.R., AND STIRZAKER, D.R. (1992). *Probability and Random Processes. Problems and Solutions*. Clarendon, Oxford.

HALDANE, J.B.S. (1945). A Labour-saving Method of Sampling. *Nature*, **155**, 49.

KEMENY, J.G., AND SNELL, J.L. (1983). *Finite Markov Chains.* Springer, New York.

MISES, R. VON (1931). *Wahrscheinlichkeitsrechnung.* Franz Deuticke, Leipzig and Wien.

MORGAN, B.J.T. (2000). *Applied Stochastic Modelling.* Arnold, London.

MOSTELLER, F. (1965, 1987). *Fifty Challenging Problems in Probability with Solutions.* (1st ed.) Addison-Wesley, Reading; (2nd ed.) Dover Publications, New York.

PÓLYA, G. (1975). Probabilities in proofreading. *American Mathematical Monthly*, **83**, 45.

ROSS, S.M. (2000). *Introduction to Probability Models.* (7th ed.). Academic Press, San Diego.

ROSS, S.M. (1996). *Stochastic Processes.* (2nd ed.). Academic Press, London.

SZÉKELY, G.J. (1986). *Paradoxes in Probability Theory and Mathematical Statistics.* Reidel, Dordrecht.

Index

Note: f. and ff. refers that the topic is discussed in next consecutive page numbers.

Problem Books in Mathematics *(continued)*

Theorems and Counterexamples in Mathematics
by *Bernard R. Gelbaum and John M.H. Olmsted*

Exercises in Integration
by *Claude George*

Algebraic Logic
by *S. G. Gindikin*

Unsolved Problems in Number Theory (Third Edition)
by *Richard K. Guy*

An Outline of Set Theory
by *James M. Henle*

Demography Through Problems
by *Nathan Keyfitz and John A. Beekman*

Theorems and Problems in Functional Analysis
by *A.A. Kirillov and A.D. Gvishiani*

Problems and Theorems in Classical Set Theory
by *Péter Komjáth and Vilmos Totik*

Exercises in Classical Ring Theory (Second Edition)
by *T.Y. Lam*

Exercises in Modules and Rings
by *T.Y. Lam*

Problem-Solving Through Problems
by *Loren C. Larson*

Winning Solutions
by *Edward Lozansky and Cecil Rosseau*

A Problem Seminar
by *Donald J. Newman*

Exercises in Number Theory
by *D.P. Parent*

**40 Puzzles and Problems in Probability
and Mathematical Statistics**
by *W. Schwarz*

Functional Equations and How to Solve Them
by *Christopher G. Small*

**Contests in Higher Mathematics:
Miklós Schweitzer Competitions 1962–1991**
by *Gábor J. Székely (editor)*